PRINCIPLES OF LANDSCAPE SCIENCE AND PHYSICAL–GEOGRAPHIC REGIONALIZATION

Principles of Landscape Science and Physical-Geographic Regionalization

A. G. ISACHENKO

Edited by John S. Massey

Assisted by N. J. Rosengren

TRANSLATED BY R. J. ZATORSKI

MELBOURNE UNIVERSITY PRESS
1973

First published 1965 by
Vysshaya Shkola, Moscow

First English language edition 1973

Printed in Australia by
Wilke and Co. Ltd, Clayton, Victoria 3168 for
Melbourne University Press, Carlton, Victoria 3053
USA and Canada: ISBS Inc., Portland, Oregon 97208
Great Britain and Europe: Angus and Robertson (UK) Ltd, London

Registered in Australia for transmission by post as a book

ISBN 0 522 84046 9
Dewey Decimal Classification Number 910.02

Foreword

This text, initially published in the Russian language in 1965, represents the first attempt to draw together philosophical, theoretical and methodological concepts of physical-geographic differentiation and present them to advanced students of physical geography. In addition, it sets forth principles for the classification and regionalization of terrain.

Geography as landscape science has been discussed at length since 1960 by various authors in articles appearing in *Soviet Geography: review and translation*. Landscape science represents that portion of physical geography which considers the differentiating effects of solar energy (exogenous forces) and the internal energy of the earth (endogenous forces) on the earth's surface. The result of the interaction of these forces is a hierarchy of individual, 'concrete' units (geocomplexes), each having a unique association of climate, geology, landform, soils, vegetation as well as human produced features. It is the aim of Soviet landscape science to analyse and classify geocomplexes in order to form a basis for planning and economic development.

Physical-geographic regionalization as part of landscape science is concerned with the delineation of individual geocomplexes of the earth's surface. Essentially it is based on an understanding of genetic factors with particular emphasis on the laws (theory) of terrain differentiation. Although there is variation in the effect of the various exogenous and endogenous forces with variation in scale of observation, both zonality and azonality occur simultaneously to render the earth's surface a mosaic of units. There is not one mosaic, however, but several. At large-scale, we see a map subdivided into relatively small, internally homogenous units, while at small-scale the units are internally heterogeneous and perhaps several thousand square kilometres in extent. From large- to small-scale, the main physical-geographic regional categories recognized by Isachenko in this text are facies (facia—singular), urochishcha (urochishche—singular), landscapes, provinces, subzones and zones. Since a landscape represents an area characterized by uniformity of both zonal and azonal conditions, it is considered the central unit in this schema and given most attention by the author.

In chapter 1 the subject matter of landscape science is defined and compared with that of general physical geography. In addition, principles of terrain differentiation and physical-geographic regionalization are set forth and briefly discussed. The history of landscape science is then treated in four sections. The first two deal with the development of landscape science ideas prior to the formalization of the discipline, while the following two consider the contributions of V. V. Dokuchayev, his students and followers, and the subsequent emergence of landscape science in the mid-1950s as a formal branch of Soviet physical geography. The final section of chapter 1 deals with the development of landscape science outside the U.S.S.R.; special reference is made to the regional concepts of geographers of the United States.

The second chapter deals in detail with the laws of terrain differentiation. Factors which contribute to zonality, especially climate, and the various manifestations of zonality in landforms, soils, vegetation and so forth are considered in the first two sections of this chapter. The existence of zonality in early geological time is recognized and its development (to the Recent) over the geological record is traced. Azonality is defined as an independent geographic principle and is considered of equal importance to zonality. Although a number of synonymous terms including 'meridional zonality', 'sectionality' and 'provinciality' have evolved, the author feels that all these may be grouped under the name 'azonality'. '. . . in the final analysis each is due to a single factor: the history of the tectonic development of our planet resulting from the action of its internal energy.' Vertical stratification, the marked variation in surface character with elevation, especially in mountain regions, is discussed in the fourth section. Although vertical stratification is primarily the result of azonal factors, the importance of latitudinal position is recognized. Azonal differentiation of plains is dealt with in the fifth section, and while in many respects it is fairly similar to that obtaining in high altitude areas, its effects are more difficult to appreciate because of broad transition areas.

The author treats zonality and azonality as two mechanisms of independent origin but operating simultaneously to form the basic physical-geographic unit, the landscape. Below the landscape level are the morphological units, facies and urochishcha, which are briefly discussed in the concluding section of this chapter.

Landscapes, urochishcha and facies are defined and analysed in chapter 3. The first four sections deal with general problems of landscape delineation and internal organization. Special reference is made to landscape content, for example geology, soils, vegetation and structure, the form resulting from the interaction of individual components. In the third and fourth sections problems associated with boundary mapping and with determining *morphological structure*—the arrangement of urochishcha and facies within a landscape—are discussed. The point is made that boundaries are not usually represented by constant changes in one component, say landform. With regard to a given landscape, two or more components, simultaneously altering, may represent a boundary. Or a significant change in one component may represent one portion of a boundary while a change in another component may represent another portion of the same landscape boundary. In the fifth and sixth sections the large-scale units, urochishcha and facies, are discussed in detail. The second half of chapter 3 is concerned with more specific problems of landscape science such as mountain landscapes, landscape geochemistry, landscape dynamics, the effects of man's activities on landscapes, and landscape classification. Of particular interest are the ninth and tenth sections dealing with landscape dynamics. Landscapes are considered as dynamic systems to which the author relates Marxist laws of dialectics.

The final chapter deals with physical-geographic regionalization. After discussing basic concepts underlying this aspect of landscape science and the history of regionalization, the author presents the various zonal and azonal regionalization schemata of Soviet geographers. Since regionalization is primarily oriented to the systematization of large areas, special emphasis is placed on small-scale units: zones, districts, etc. Problems associated with the combination of zonal and azonal schemata in 'a single subordination series' are discussed in the following section. The author proposes a biserial system of taxonomic units as a solution to these problems. 'The biserial system is based on an objective treatment of both the primary series—zonal and azonal—which merge in the landscape and which are associated on intermediate levels by productive or linking units.' Because of the individuality of mountain areas their regionalization is treated separately in the seventh section. This section is followed by consideration of the relationship between landscape regionalization and the regionalization of individual geographic components. In the author's opinion 'There are no theoretical grounds for the rejection of the correspondence between landscape regionalization and the branch-types of regionalization. Neither should we admit a simplistic concept that types of natural regionalization fully coincide with one another at every stage of classification. *Correspondence* and *coincidence* are two different notions. Were it not so, the need for branch regionalization would disappear.' The concluding sections deal with methods of regionalization and regional descriptions. The author's attitude to complimentary use of regionalization 'from above' and 'from below' closely corresponds with that of Hartshorne reported in his 1939 paper in the *Annals of the Association of American Geographers.*

The transliteration system used in this translation from the Russian language was that adopted by the American Geographical Society in its publication, *Soviet Geography: review and translation.* Occasionally it was necessary to alter translated spellings of place names in order to maintain correspondence with names found on some commonly available maps of the U.S.S.R.

Western geographers have contributed to the development of theory and methods of classification and regionalization of the earth's surface. Professor Herbertson's classic paper, 'The major natural regions: an essay in systematic geography' (1905), although criticized after its presentation, initiated considerable interest in regional geography in Great Britain. Herbertson's work was followed by contributions by Roxby, Unstead, Crowe, Dickinson, Robinson and others dealing with many aspects of areal classification and regionalization, but concentrating primarily on methodology. For the most part these geographers were interested in relatively large, heterogeneous areas, which reflected in some way a unity of physical and cultural conditions. The work of British geographers was paralleled by that of the Americans including Joerg, Fenneman, Sauer, Finch, Platt, Jones, Thornthwaite, Whittlesey, Hall, Hartshorne and James, many of whose articles appeared in the *Annals of the Association of American Geographers* between 1915 and 1940. While some of these workers concentrated on producing broad, multi-aspect classifications and regionalizations, others sought to limit the criteria of area delineation. These latter efforts resulted in the publication, for example, of climatic, agricultural and physiographic classifications and regionalizations for large areas of North America. By the early 1950s there had been considerable refinement of the regional concept and the publication of several major review articles, notably those of James, Whittlesey, Linton, and Kimble. Concepts of regional geography had been refined and a variety of types of regions, formal or uniform, nodal or functional, etc., were recognized. Meanwhile Kimble's attack on 'traditional' regionalization heralded a decline of interest in the 'catch-all', encyclopaedic style of regional geography.

Simultaneously, however, new interest in regions and regionalization was stirred by the need to understand complex relationships of socio-economic variables in areas of developed and developing nations. Out of this need, regional science and the Regional Science Association, 'devoted to the free exchange of ideas and viewpoints with the objective of fostering the development of theory and method in regional analysis and related spatial and areal studies', were born. The availability of high speed data-processing equipment in about 1950 added a new dimension to the analytical capabilities of social scientists and geographers. It was not long before the tools of multivariate statistical analysis, some similar to those employed for social area regionalization by Hagood in the early 1940s, were being applied by geographers in a wide variety of regional geographic situations. While much of this work was done in the United States by Berry, Zobler, Mackay and others, most departments of geography now have at least one staff member dabbling in regionalization through quantification. The marriage of mathematics and regional concepts is considered a viable alternative to the pre-quantitative methods of regional delineation.

While research emanating from universities has had an important influence on physical-geographic classification and regionalization, most of the contributions to this aspect of regional geography have come in recent times from individuals associated with government physical-resource survey organizations. The land type concept evolved from the work of Jethro Veatch and other members of the Land Economic Survey in northern Michigan in the 1920s. At first the land types were envisaged as summary devices representing the combination on one map of geology, soil and vegetation conditions of each of the counties studied by the survey. Later, land types were identified and delineated as composite units. Early in the 1930s the Depression and consequent lack of funds for state land resource projects resulted in the premature demise of the Land Economic Survey, but this did not happen before the ideas had received moderate publicity and a number of geographers and geography students had received training in survey methods and concepts. A full report of the activities of the Land Economic Survey is found in Davis, *A Study of the Land Type* (1969).

After World War II certain British Commonwealth governments turned their interest to undeveloped areas of Australia, Africa and New Guinea. Many of these areas had not been explored let alone mapped, and survey organizations such as the Commonwealth Scientific and Industrial Research Organization (CSIRO) Division of Land Research were established to carry out the primary survey work. Unlike the Land Economic Survey, CSIRO sought initially to map composite units, and the land unit and land system concepts were born. A full report of the activities of the Division of Land Research is found in 'Methodology of integrated surveys' (Christian and Stewart, 1968). The methods and concepts of CSIRO have had an important influence on applied physical geography. They have been adapted, and in some cases altered quite significantly, to survey situations in many parts of the world. This work continues today forming a basis for planning in newly developing nations.

Much more could be said about the contributions of survey organizations and academic geographers, sociologists, economists, etc., to the development of regional geography. The point I have tried to make, which is similar to that expressed by R. E. Dickinson in *Regional Ecology*, is that regional geography has been of great interest to geographers and still remains an important focus of geographic research. It is partly for these reasons that this book is published in the English language.

The text should appeal to academic geographers interested in tertiary education in geography in the U.S.S.R. It should also appeal to those involved in physical-geographic classification and regionalization. Since it is a textbook the material covered should be of interest to advanced students of physical geography. Furthermore, it has been prepared as much for non-geographically trained readers as for those experienced in the field. It is not a final statement of Soviet landscape science and physical-geographic regionalization; rather, it is a summary and interpretation of the stage to which this aspect of Soviet physical geography had advanced by 1965.

It was necessary to alter the literal translation in many instances in order to make the text easier to read; at all times considerable attention was paid to maintaining the original ideas of the author. I accept full responsibility for any discrepancies between the Russian and English texts.

Several people have contributed to the publication of this book and to them I offer my thanks. Richard Zatorski, beyond his duty as translator, provided constant encouragement and invaluable guidance. Sue Zatorski, Natalie Trunoff and Liz Johanson typed the manuscript and Neville Rosengren assisted greatly with editing. Dorothy Prescott did a splendid job with the indexing. My wife, Kathie, has infinite patience. Finally, let me thank A. G. Isachenko for permission to publish his book in English. It is my sincere hope that we shall one day meet.

Melbourne   *John S. Massey*
1973

Preface to English Edition

Contemporary geography constitutes a vast complex or system of sciences. In Russia, since the end of the nineteenth century, the differentiation of geography was accompanied by the emergence of an independent, synthetic direction—the science of landscape. In the U.S.S.R. this new direction has achieved widespread development and today occupies a central position in the entire system of geographic sciences.

The geographic or natural territorial complex or, rather, the geocomplex,* is a fundamental concept in Soviet geography. All geographic sciences in one way or another address themselves to the study of geocomplexes. Only for physical geography, however, are these systems the proper object of investigation. At the same time general physical geography is concerned with the study of the geocomplex of the entire planet, the so-called geographic or landscape envelope of the earth (which we designate as an epigeosphere). Geocomplexes on the regional and local level (zones, provinces, landscapes, facies, etc.), which jointly form the 'geographic mosaic' of the earth's surface, are the subject of landscape science. This book is devoted to the exposition of the fundamental principles of landscape science.

It is worth noting that following its publication in the U.S.S.R. in 1965, a large number of monographs concerning various aspects of landscape science have been added to Soviet geographic literature. Recently the systems approach has been adapted to the study of landscapes. A. D. Armand,† V. S. Preobrazhenski and V. B. Sochava have been developing the concept of landscapes and facies as self-

---

* 'Geocomplex' was originally translated as 'geosystem', but edited to 'geocomplex' to preserve the conformity of the entire text. The use of geosystems is undoubtedly an attempt to keep pace with current jargon. (Saushkin and Smirnov, 1970)
† Armand (1969)

regulating information systems. Attempts are being made to develop mathematical models of geocomplexes. In the area of landscape methodology we find a tendency towards a broader utilization of mathematical methods. Field-station investigations of the dynamics of geocomplexes, employing geophysical and geochemical methods, are also under development. A special mention should be made of the various applied branches of landscape science currently under development; these are concerned with territorial organizations for regional planning, the establishment of recreational zones, agricultural soil classification, and the medical-geographic and engineering evaluation of lands.

This book is not concerned with the problems of applied landscape science; it is an attempt to provide a systematic exposition of the theoretical foundations underlying this branch of physical geography. Simultaneously it constitutes the first and virtually the only university-level textbook on landscape science. The next two or three years should also see the compilation of a textbook on applied landscape science.

It is well known that in English-speaking countries geography has also had an illustrious history, and although the study of landscape as a theoretical discipline did not attain a high level of development in these countries, the last decade saw the emergence of applied studies in soil classification, regional planning and silviculture, of concepts and methods very close to those accepted in Soviet landscape science. This is true, in particular, of the soil mapping in the mid-west of the U.S.A., and subsequently in the Tennessee Valley during the thirties, employing a method which Hudson designated as the 'unit area method'.* During 1949–51 similar investigations were carried out by North-Western University in Puerto Rico. Before World War II, L. D. Stamp developed principles for the classification of soils which formed the basis for the soil map of the United Kingdom.

From the point of view of landscape science, great interest attaches to the concept of 'site', introduced in 1931 by R. Bourne. This is identical with the Soviet geographic concept, facia, constituting an elementary geocomplex. There are many areas of contact between

* A forerunner of this work was that of J. O. Veatch and other members of the Michigan Land Economic Survey. Although the influence of this group on 'land students elsewhere in the country and abroad' is questionable, it cannot be denied that many were well aware of the survey and the concepts it had derived and applied. (Davis, 1969, pp. 35, 41) Although such methods are no longer in widespread use (with the exception of the state of California), new general-purpose survey methods employing rigorous sampling procedures and multi-variate statistical analysis have been adopted recently. (Tennessee Valley Authority, 1965)

landscape science and the study of ecosystems, in particular forest
ecosystems as developed by Canadian silviculturists.*

The closest in character, however, to the landscape investigations
of Soviet geographers are the soil surveys carried out by CSIRO in
Australia, whose methods are described by C. S. Christian and
G. A. Stewart.†

These are only a few examples, but they are sufficient to indicate
the existence of mutual interests in the area of geographic theory and
method. It is hoped that the publication of this book in an English
translation will assist our foreign colleagues in familiarizing themselves with the theoretical foundation of Soviet landscape science. A
friendly critique will be welcome.

Leningrad State University A. G. Isachenko
July 1971

* Hills (1960; 1961, pp. 1-204); Rowe (1962, pp. 420-32)
† (1952) This is the first report of the Land Research Series prepared by the
Division of Land Research. Other reports in the series, which now numbers
almost thirty and covers areas in Queensland, the Northern Territory, Western
Australia and Papua New Guinea, are also available.

Preface to Russian Edition

The study of the principles of landscape science is an important link in the university-level training of expert physical geographers. In recent years the literature dealing with landscape science has been enriched by many valuable contributions. Among these, however, there are no textbooks. The present book constitutes the first attempt to elaborate, in a systematic fashion, the principles of landscape science in a form accessible to advanced students already familiar with general and regional physical-geographic disciplines. The text is designed in conjunction with the curriculum of a course of the same name given at the Department of Physical Geography in the Leningrad State University. The core of the book is a lecture course given by the author over a number of years at that university and also at the Sun Yat Sen University in Guanchow and the University of Peking.

The bibliography includes only a small number of the most important titles. Some additional sources dealing with material in the individual chapters are quoted in the notes.

A. G. I.

Contents

Figures

Note on Arrangement

In the original text there are both footnotes and a bibliography. In this edition the author's references and comments are indicated in the text by a superior figure and are grouped by chapter at the end of the book. The editor's notes to the text, primarily of a comparative nature, appear as footnotes.

J. M.

1

The Subject, Scope and History of Landscape Science

The subject and scope of landscape science

Landscape science is a rapidly developing scientific discipline which constitutes an extremely important branch of contemporary physical geography.

Physical geography studies the so-called geographic envelope of the earth, its composition, structure, the laws of its development and the laws of terrain regionalization.[1] The term 'geographic envelope' denotes the most complex part of our planet including the troposphere (the lower levels of the atmosphere up to an altitude of 8 to 16 km), the hydrosphere (to a maximum depth of approximately 11 km) and the lithosphere (4 to 5 km thick), as well as the sum total of the organisms inhabiting these three geospheres.

The geographic envelope, within the limits indicated, constitutes an integrated material system, qualitatively* different from the remaining parts of the globe. Its internal unity is imparted by the continuous exchange of energy and matter which takes place between the individual components† of the envelope. A specific feature of the geographic envelope is the fact that in no other areas of the globe does matter exist simultaneously and stably in the solid, liquid and gaseous states; the substances in all these three states interpenetrate and interact with one another. The second feature of the geographic envelope arises from the fact that numerous processes, due to the absorption and transformation of solar radiation energy, take place within its limits; these processes enter into extremely complex relationships with the various manifestations of the earth's internal energy. The qualitative distinctness of the geographic envelope is determined, further-

* 'Qualitatively' is used here in the sense of the system representing more than just the simple sum of its parts.
† 'Components' is used throughout the text to refer to geology, soils, vegetation, etc.

more, by its role as a sphere in which organisms are created and developed, and in which they carry on their various activities. Finally, the geographic envelope affords human society a stage for biological life and its change-generating productive functions.

In geographic literature the terms 'geographic sphere', 'landscape sphere' and 'landscape envelope' are employed as synonyms for our term 'geographic envelope'. Some geographers think that the term 'geographic envelope' (or 'sphere') is redundant since it repeats the name of the science itself ('geographic envelope': the subject of geography) and does not expose the sense of the subject. On the other hand, the terms 'landscape envelope' and 'landscape sphere' must be viewed as equally indistinct, since they are sometimes used in a different sense: to denote a narrow layer, some tens of metres thick, directly underlying the surface of dry land. I. M. Zabyelin proposed 'biogenosphere' as a replacement for 'geographic envelope' so as to stress the fact that the organic world has originated in this section of the earth. Such a designation, however, emphasizes only one, even though very important, characteristic of the geographic envelope and seems somewhat biased. An acceptable alternative is the term 'epigeosphere', i.e. the external envelope of the earth, which may be used as a synonym for 'geographic envelope' on the analogy of our use of 'atmosphere' as a synonym for the 'air envelope', 'hydrosphere' for the 'water envelope', etc.

The most complex parts of the geographic envelope are where the lithosphere, hydrosphere and atmosphere meet and interact directly, i.e. on the surface of dry land (more precisely, the surface horizon of the lithosphere, together with the ground level of the atmosphere, surface and ground water); in the upper layer of the earth's oceans; and on the ocean bottom. It is at these interfaces of the geographic envelope that the maximum concentration of biological matter is observed.

An important result of the interaction of the lithosphere, hydrosphere and atmosphere is the complex spatial differentiation manifested in the great variety of observed combinations of such geographic components as rocks, bodies of water, soils, etc. In the course of the development of the geographic envelope its components formed on the earth's surface regular and territorially limited entities called *geographic* or *natural territorial complexes*, referred to as *geocomplexes*. In the following pages we shall discuss only the geocomplexes of dry land, which are characterized by maximum complexity and diversity; their components are the surface rocks and loose deposits with their inherent relief, the ground layer of the atmosphere with its climatic peculiarities, surface and ground water, soils, and concentrations of plant and animal life. *All these natural components develop within a single material system; the realization of this fact underlies the concept of the geocomplex.* It follows that the internal unity of an individual geocomplex, whether large or small (e.g. the entire tayga zone or a small, swampy lowland), is analogous to that of the geographic envelope; it arises from the interpenetration, interaction and simultaneous development of all its components. The internal unity of the geographic envelope finds its concrete expression in every geocomplex.

The discussion to this point leads necessarily to the conclusion that the geographic envelope could conceivably be studied in two ways: as an entity, and as a set of geocomplexes. The former constitutes the aim of general geography,* the latter that of landscape science.

The scope of general geography, regarded as an independent subject, includes the following: (1) the study of the geographic envelope as part of the earth and the universe, the external influences of outer space and the planets, and the effects of tectonic activity and their transformation within the geographic envelope; (2) the study of energy relationships in geographic processes; (3) the study of the geographic envelope, its chemical and physical composition and its cycles; (4) the study of the structure of the geographic envelope, its components and the relationships between them; and (5) the study of the laws governing the development of the geographic envelope, its rhythm and its successive variations.

Another aspect of physical geography is comprised of the study of the laws of the geographic differentiation of terrain; it includes, in particular, the study of physical-geographic zones. This aspect represents a transition to landscape science. Naturally a sharp boundary cannot be drawn between general geography and landscape science; both deal with closely associated aspects of a single science. Knowledge of the laws underlying general geography is a necessary prerequisite for the study of landscape science; at the same time landscape science is

* This is in keeping with the division of geography devised by Varenius in 1650. "In contrast with most of the writers of his time who limited geography largely to bare description of the several countries—including therein much of their political constitutions—Varenius divided the field into 'general or universal geography' and 'special geography' or 'chorography'.
The terms which Varenius used, 'general' and 'special' geography, later became the standard terms in Europe for these two aspects of the field, though many later writers have found them unsatisfactory. As we will note in a later section, the frequent use by German writers of the adjective '*systematisch*' in describing 'general geography' supports the common use in this country of the term 'systematic geography'. The term 'special geography' was largely replaced in German literature by '*Landerkunde*' which in spite of obvious disadvantages is commonly favoured over non-Germanic terms, 'special', or—the term now nearly universal outside Germany—'regional geography' ". (Hartshorne, 1939, pp. 217-18)

a natural continuation of general geography, comprising that portion of the science of geography in which the laws of general geography find concrete application to the task of explaining local geographic peculiarities. This accounts for the great practical importance and broad scope of landscape science. It is precisely this area of geographic science which bears directly on the problems of complex utilization, protection and regeneration of natural resources of individual zones, districts and regions.*

The subject of landscape science is the study of geocomplexes, their morphological structure,† development and distribution. In other words, *landscape science is an area of physical geography which studies the territorial differentiation of the geographic envelope.*

The surface of the earth consists of geocomplexes of different order, ranging from the smallest and relatively simple to the largest and most complex. These geocomplexes interlink with one another to form a system of geographic units subordinate to the geographic envelope.

Many Soviet geographers adhere to the view that in a system of physical-geographic terrain classification, one stage should be distinguished and treated as basic. Such a basic stage is the *landscape*, which gave the name to landscape science. Hence the study of landscape as the fundamental object in the physical-geographic investigation of terrain constitutes the nucleus of landscape science. A view has also been expressed[2] that landscape science should restrict itself to the study of landscapes and should exclude problems having to do with the study and systematic classification of more intricate complexes (landscape zones, districts), i.e. it should avoid what is called physical-geographic regionalization. However, one can hardly juxtapose the study of the landscape and physical-geographic regionalization, since the landscape itself constitutes a unit of regionalization and all the higher physical-geographic units represent the territorial amalgamations of landscapes; these units are known and characterized by their landscapes. Physical-geographic regionalization constitutes a special kind of systematic classification of landscapes, and for this reason it is treated as an area of landscape science.

Landscapes, in turn, may be subdivided into simpler geocomplexes, i.e. urochishcha and facies, which are regarded as the morphological units of a landscape. Their study is undertaken by one section of landscape science called landscape morphology.

* 'Zones', 'districts' and 'regions', when considered individuals rather than types, are regional units. These are discussed in detail in chapter 4.
† 'Morphological structure' refers to the relationships between the various units, facies and urochishcha, which constitute a landscape.

Geocomplexes are examined both as individual entities and as members of a typological schema. This means that from the point of view of theory and practice, interest is focused on the one hand on every individual geocomplex (e.g. the Russian plain, the forest-steppe of the Russian plain, the Polesye, the Dneper floodplain, etc.), and on the other hand common features, general (typological) characteristics, unifying the various complexes are discerned. In other words, an attempt is made to reduce the diversity of the individual geocomplexes of a given category (e.g. landscapes) to a number of species, classes and types.

This gives rise to two separate approaches to the study of geocomplexes: the individual and the typological approaches. Obviously the individual approach gains priority when we study intricate complexes or complexes of higher rank, i.e. landscape and above, which are designated as *regional physical-geographic units*, whereas the typological approach dominates the study of simple geocomplexes, i.e. the *morphological units* of landscape.

It must be emphasized that in nature there exist objectively only concrete, individual geocomplexes of different rank.* When these are classified, i.e. when they are assigned on the basis of their qualitative similarities into species, classes, etc., a complex system of *typological units* is created. Every typological unit, for example a landscape type or an urochishche type, results from a theoretical generalization. When generalizing this way, the inherent, individual properties of the objects under study are disregarded and only those properties which are common to them all are abstracted. It is wrong, therefore, to designate, as is sometimes done, simple geocomplexes as typological units on the grounds that we are dealing more frequently with types or species than with concrete complexes. The concept of type can only arise from the differentiation and comparison of concrete geocomplexes, whether dealing with facia, landscape, or any other unit. It follows that the knowledge of individual units is primary, whilst typological concepts are secondary and derived.

Every category of geocomplexes is classified separately, and for this reason landscape science includes not one but many independent

* A number of arguments concerning the *existence* of physical units over the surface of the earth have been presented in both Soviet and Western literature. (Crowe, 1938; Hartshorne, 1939, pp. 426-60; Szava-Kovats, 1966; Yefremov, 1961) Discussion is not confined to the consideration of composite units, and it has been considered at length with regard to continuum concepts of vegetation. (McIntosh, 1967) Although the problem is not resolved, it does seem to be receiving considerably less attention in the literature than it was once accorded. For the most part, Western geographers deny the existence of units but utilize a regional framework for making generalizations about an area. The question of existence is discussed at the end of this chapter.

classifications, e.g. for facies, for urochishcha, for landscapes, etc. For the classification of higher regional units (provinces, zones) the problem is not very serious: the number of such units in nature is relatively small and reduces with the rise in rank. Concurrently, the number of idiosyncratic features continues to increase in each higher unit; this increase in individuality in the end eliminates the need for detailed classification.

Systematic relationships between the principal categories of physical-geographic regionalization may be illustrated by a simple schema. (Fig. 1)

The study of geocomplexes proceeds by various methods; the most important of these is field investigation, accompanied by landscape mapping. The methodological problems in landscape investigations and cartography are a subject of a separate study and are not discussed here.[3]

In this book landscape science is treated in the broad sense, and is divided into three major chapters: the study of the laws of physical-geographic differentiation; the study of landscape proper, i.e. landscape science in a narrow sense, with sections devoted to structure, dynamics, morphology, landscape classification, etc.; and physical-geographic regionalization.

Within this outline, landscape science practically coincides with *regional physical geography*. Regional physical geography is still often represented as a purely descriptive discipline aimed at the derivation of physical-geographic descriptions of individual territories. Such a discipline, however, is completely outmoded. Descriptions of course constitute the unavoidable burden of every science, but are not the basis of the science itself. Regional physical geography can only be transformed from a descriptive branch of geography into a scientific, theoretical discipline against the background of a theory of landscape. Only then would conditions exist for a reconstruction of the content of regional physical-geographic descriptions. Meanwhile such descriptions continue to be constructed, according to the traditional pattern, as component lists, hardly ever accompanied by an analysis of the interactions and the laws of physical-geographic differentiation.

To produce a meaningful physical-geographic description, regardless of whether the terrain to which it applies is delineated by natural or political boundaries, is not to produce an 'item-by-item' description (e.g. to describe separately the climate, relief, etc.), but is to give an account of the geocomplexes—the natural system of terrain classification. Furthermore, a regional physical-geographic description should adequately reflect the laws affecting geocomplexes; it must reveal their origin, history and future development trends, current processes,

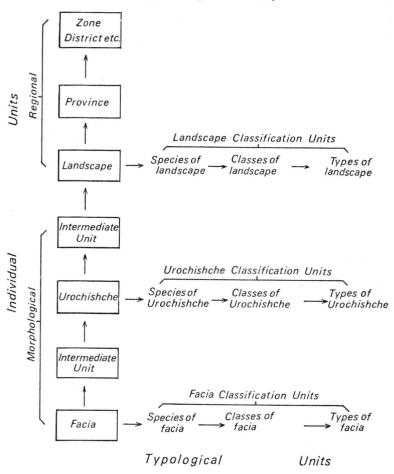

Fig. 1   System of units of physical geography

internal diversity (morphology), mutual interactions, as well as the effect of human society on all these factors. In other words, regional physical-geographic description must conform to the current theoretical ideas about the laws of the formation, origin and structure of geocomplexes.

In conclusion, the problem of the relationship between landscape science and the so-called regional geography of nations must be discussed. The regional geography of nations is a supplementary scientific discipline; its task is the compilation of significant data on the various sections of the earth's surface, usually within political or administrative boundaries. It includes discussions on natural conditions as well as population, culture, national economy, politics, history and art. The practical utility of combining in a single source diverse data about any one country is quite obvious, but the regional geography of nations does not of course constitute in itself an independent science. The material for its descriptions is borrowed from various sciences, one source of which is landscape science. But this should not lead to the identification of landscape science with the regional geography of nations.

The beginnings of landscape science

The great Russian scientist V. V. Dokuchayev (1846–1903) is considered to be the father of landscape science. Dokuchayev's work was accomplished mainly at the end of the nineteenth century, yet landscape science became established in Russia as an independent discipline only after the October Revolution. The origins of landscape science are very much deeper; its roots are in the practical experience of the people, and its entire development has always been associated with the problems of the national economy. The concepts of landscape science, on the other hand, have their origin in the development of natural science, commencing approximately in the mid-eighteenth century.

Since time immemorial, daily existence continuously confronted man with the need to distinguish natural areas, areas distinguished from one another by their natural resources and methods of cultivation. Practical necessity prompted individual nations, long before scientific ideas about landscape had emerged, to develop empirical concepts about geocomplexes. The uniqueness of certain localities was reflected in such place names as Polesye, Myeshchyora, Kargopol, Zhiguli, etc.

Certain doubtlessly synthetic, folk-originated designations have become firmly established as names for various landscape types (e.g. tundra, tayga, steppe, desert, etc.). The idea of tayga, for example, embraces not only the plant life but all the typical natural features including the relatively severe climate with winter snowfalls and poor podsolic soils together with the characteristic animals. The same could be said about the tundra, steppe and desert.

In folk speech there is a genuine taxonomy of geocomplexes. Paralleling the broad classification shown in the above examples, simpler territorial units, designated in Russian as 'urochishcha', have long been distinguished. Urochishcha comprise individual forests, marshlands, salt-marsh depressions, gullies, etc.

The geographic ideas developed by the indigenous population as a result of prolonged experience played a considerable role in the development of the science of geography. The great Russian geographers, especially V. V. Dokuchayev and his successors, repeatedly turned to this source. G. F. Morozov urged the people to view the forests and the land on which they lived as a specific entity—a 'geographic individual'.

The need to account fully and adequately for the local natural conditions and resources in economic management inevitably generated interest in terrain regionalization. In Russia this interest dates to the eighteenth century, becoming especially intense towards the end of that century as a consequence of the rapid rise of capitalism, the development of industry, and the intensified economic specialization of individual areas. Numerous publications concerned with the natural regionalization of Russia played an important role in the development of concepts in landscape science. This is despite the fact that over a period of a century these publications bore an empirical character and were not based on any underlying theoretical foundation. Landscape science had to provide such a foundation, and sooner or later physical scientists and geographers would have had to elaborate a scientific concept of landscape. The path towards that concept, however, was long and difficult to negotiate.

Every scientific theory draws on the entire prior development of science and emerges only when certain basic prerequisites have been fulfilled. Before scientific landscape study could emerge, it was necessary, first of all, to build up a sufficiently accurate concept of the shape, the dimensions and the movements of our planet and, later, a concept of the relationships between land and sea, the principal features of relief, climate, water, soil, plant and animal life. And, finally, the physical, chemical and biological laws governing natural phenomena on the earth had to be determined. In other words, the necessary prerequisites for the development of landscape science and scientific geography in general include a certain minimum of factual material, descriptive and cartographic, plus a sufficiently high level of development within the natural sciences. Such conditions did not exist until the end of the nineteenth century.

The rapid broadening of the geographic horizon and the accumulation of knowledge about the nature of the various components of the earth's crust began at the end of the fifteenth century, during the Renaissance and the period of great geographic discoveries. The first branches of knowledge to undergo relative development were those which were associated with the study of the simplest forms of motion, in particular, in the words of F. Engels, the elementary natural science: the mechanics of solids. Later, during the seventeenth and eighteenth centuries, the mechanics of liquids and gases, molecular physics, chemistry, etc. developed. Biology became an independent science only during the nineteenth century.

The study of landscape could not arise without a prior analytical stage in the development of geography which led to a deeper understanding of its constituent disciplines, e.g. climatology, geomorphology, etc. Particularly important for landscape science are biogeography and soil science. Owing to the peculiar specificity of the objects they study, biogeographers and soil scientists were the first to encounter the system of mutual relationships between living and non-living nature, moving closer than any other group of specialists towards a geographic synthesis. Biogeography and soil science, however, began to emerge into independent branches of science only towards the end of the nineteenth century, i.e. during the period which gave rise to contemporary physical geography.

Although geography is a very ancient science, it has over the last few decades undergone a period of regeneration. During its entire history, comprising some thousands of years, the task of geography had been to broaden the spatial range of vision and to accumulate facts. This old 'data-gathering' geography burnt itself out, as it were, towards the end of the nineteenth century, only to rise again on a new foundation, resting on those scientific disciplines which developed within it. In the words of the leading Russian geographer A. N. Krasnov, contemporary geography is an offspring of the natural sciences to which it once gave birth. This, of course, does not mean that the new and old geographies are separated by a deep chasm. Contemporary landscape science concepts did not emerge suddenly; their seeds can sometimes be detected in the history of science. For this reason, if we are to achieve a correct understanding of the theoretical foundations of landscape science, we must gain an appreciation of the unbroken bond linking the new geographic concepts with the past history of natural science and geography.

The development of geographic ideas
prior to the emergence of landscape science

Elementary geographic knowledge, i.e. the simplest ideas about the immediate natural environment, existed in primitive human society. Several thousand years ago the ancient cultures of the East made use of primitive maps and geographic descriptions; these were compiled for practical purposes, cadestral (land surveys for tax-gathering purposes), commercial, and others. It is true, of course, that geography can only be viewed as a separate branch of knowledge during that period in a most tentative way. The branches of science were not differentiated in antiquity and formed part of philosophical enquiry.

Ancient geography reached its peak in Greece during the Classical and the Hellenic periods (sixth to second century B.C.). Even at that time it is possible to detect the beginnings of two major directions in geography: general geography and the regional geography of nations. These are directions which persist even today. Both developed quite independently of one another. Concepts of general geography, comprising the study of the earth's shape, the nature of various phenomena (meteorological, seismic) were not only associated with philosophical ideas but also with astronomy and mathematics. In the beginning these ideas did not emerge from direct observations and were not intellectual in character; they constituted natural-philosophic guesswork, often naive and displaying a touch of fantasy, e.g. the attempt by Anaximenes of Miletus (sixth century B.C.) to attribute earthquakes to the cracking of the earth during drought.

The philosophers of the Pythagorean school (sixth to fifth century B.C.) were apparently the first to put forward, on purely intellectual grounds, the concept of the sphericity of the earth. The later philosophers of the Eleatic school, for example Parmenides (fifth century B.C.), developed on the basis of this concept ideas concerning climatic zonality. According to these only the central (temperate) zones of each hemisphere were inhabitable; between them lay a 'scorched' median (equatorial) zone, while 'solid-frozen' zones constituted the poles. This idea proved extremely viable; it was introduced on a map of 'climates' drawn by the fifth-century philosopher Macrobius, and enjoyed considerable popularity during the Middle Ages.

Aristotle (384–322 B.C.) gave the first proof of the earth's sphericity, thereby imparting a scientific character to the idea. At the same time, however, he placed the earth in the centre of the universe; the Christian Church accepted Aristotle's authority and the geocentric system reigned for some time. Aristotle was also responsible for a systematic description of phenomena in the solid, atmospheric and water envelopes of the earth. Aristotle's follower, Dicaearchus (326–296 B.C.) introduced the idea of equatorial and polar circles; he also tried to measure the height of mountains. Eratosthenes (276–194 B.C.) carried out the first relatively accurate determination of the size of the earth employing methods and data of astronomy and geometry.

The development to which is given the name, 'the regional geography of nations', had in ancient geography a purely descriptive character and was closely associated with history. At the source of these descriptions were voyages and military marches. The descriptions related principally to the historical, ethnographic or economic peculiarities of various nations; natural conditions received relatively little attention and reliable information alternated with fantasy. This period was characterized by attempts to link cultural and economic features with natural conditions, mainly climate. The most important representatives of the primitive regional geography of nations were Herodotus (485–425 B.C.) and Strabo (63 B.C.?–24 A.D.?).

The history of geography has always been associated, and especially in antiquity, with the development of cartography. The first attempt to portray the entire known world is credited to Anaximander (611–546 B.C.). Eratosthenes (276–194 B.C.) was the first to use a very simple cartographic projection. Claudius Ptolemaeus (*c.* 90–168 A.D.) was especially instrumental in the development of cartography. According to Ptolemy, geography constitutes 'a linear representation of that part of the earth which is currently known'. He designated a detailed characterization of individual nations 'chorography' and compiled an atlas containing twenty-seven maps. The size of his earth, however, turned out to be too small, and the size of his continents correspondingly too large.

The emergence of feudalism in Europe led to the predominance of religious philosophy and superstition, and ultimately to the stagnation of science. The weakening of commercial ties during the early Middle Ages (fifth to eleventh centuries) hindered the broadening of geographical knowledge; Greco-Roman science and philosophy were virtually forgotten, cosmographic and geographic ideas were subordinated to biblical studies; the idea of the earth's sphericity was rejected. Maps of that period are primitive; they include representations of Paradise, imaginary nations, fantastic citizenry, and monstrous freaks.

Scientific geography continued at a much higher level among the Arabs and among eastern nations. The Islamic state during the seventh to fourteenth centuries included part of the ancient cultures of the Middle East and Asia Minor. The Arabs studied the Greek heritage and borrowed a good deal from other eastern nations. Flourishing commerce assisted in extending geographic knowledge and intensified interest in descriptions compiled by voyagers. The later Middle Ages in western Europe (twelfth to fourteenth centuries) were characterized by the development of trade and commerce, further sea voyages, and the growth of cities. Geographic knowledge was again extended, largely due to the Crusades and through discussion with the Arabs. Through the Arabs the Europeans rediscovered the authors of antiquity including Aristotle and his teachings about the sphericity of the earth. New geographic ideas based on fact began slowly to appear among the ruling bible-based ideas. Thus, Albertus Magnus, called Albert the Great, (1193?–1280) tried to combine with the ancient teaching about climatic zones certain features of the animal world and of human populations. He rejected the view that the equatorial and polar zones were uninhabitable; a mention of perpetual snow is found in his writings.

The renaissance of science began with the imitation of antiquity. Ptolemy, whose texts were republished in many editions after 1475, regained his authority in the field of geography. The excessively stretched-out shape which Ptolemy assigned to Eurasia was the principal factor in the emergence of the idea that the shortest path to India and China was across the Atlantic Ocean.

The end of the fifteenth and the beginning of the sixteenth century is known in history as the era of great geographic discovery. These discoveries 'made purely on behalf of gain and, therefore, in the last resort in the interests of production'[4] extended greatly the limits of the known world. The voyages of Magellan in 1519–22 provided the ultimate proof of the earth's sphericity. New discoveries made it possible to refine the ancient knowledge of climatic zones; new data were obtained about wind zones of the oceans (the doldrums, trade winds, westerlies). Numerous corrections were made to earlier ideas about the relationships between oceans and continents.

These great geographic discoveries led to substantial economic progress and initiated the original process of capital accumulation in western European countries. The first voyage of Columbus in 1492 and Vasco da Gama's discovery of the sea passage to India in 1498 gave impetus to the colonial expansion of Spain and Portugal leading to the conquest of the Americas and of the East Indies. The discovery and colonial conquest of new lands continued, but the initiative passed in the second half of the sixteenth century to England, France and the Netherlands. From the end of the sixteenth to the middle of the seventeenth century the most outstanding geographic discoveries were made in northern Asia by the Russians.

When eventually the constraints of theology and of the authority of the ancients were laid aside, the accumulation of new natural-scientific material created the necessary conditions for a rapid development of natural science. The initial period in the history of contemporary natural science began, according to F. Engels,[5] during the Renaissance and ended with Newton (1642–1727) and Linnaeus (1707–78). During this period the heliocentric system postulated by Copernicus in 1543 and developed by Bruno, Kepler and Galileo was verified; the science of mechanics was created. The philosophy of this period was characterized by the development of materialistic concepts, particularly those of Francis Bacon and Rene Descartes.

During the middle of the seventeenth century the telescope, the thermometer and the barometer were invented, although the first consistent series of meteorological observations using these instruments was not made until the beginning of the eighteenth century.

The progress of mechanics and optics led to improvements in the methods of topographic survey, e.g. plane tabling, triangulation, barometric levelling, etc. Systematic topographic surveying was being carried out in some countries as early as the second half of the sixteenth century. At that time new and original atlases of the known surface of the earth were published in the Netherlands, the *Theatrum Orbis Terrarum* of Ortelius in 1570 and Mercator's *Atlas* in 1595. During the second half of the seventeenth century the methods used in astronomy and geodesy were improved, the first mapping surveys were carried out, and the dimensions and shape of the earth were determined more precisely.

The progress of natural science and of cartography, however, was still poorly reflected in the development of geographic ideas. Engels has shown that during the period under discussion most branches of natural science including physics, chemistry, geology and biology either existed in an embryonic state or had only commenced their development.

> There could as yet be hardly any talk of the comparison of the various forms of life, of the investigation of their geographical distribution and their climatic, etc., living conditions. Here only botany and zoology arrived at an approximate completion owing to Linnaeus.[6]

> 'But what especially characterizes this period is the elaboration of a peculiar general outlook, in which the central point is the view of the *absolute immutability of nature*. In whatever way nature itself might have come into being, once present it remained as it was as long as it continued to exist . . . The 'five continents' of the present day had always existed, and they had always had the same mountains, valleys, and rivers, the same climate, and the same flora and fauna, except in so far as change or cultivation had taken place at the hand of man . . . In contrast to the history of mankind, which develops in time, there was ascribed to the history of nature only an unfolding in space.[7]

A substantial contribution to geography, mainly concerning plant and animal distributions, resulted from observations of certain travellers. As early as 1495 Cardinal Bembo described the vertical stratification of plant life on Mount Etna. After the mid-seventeenth century, accounts of 'natural-scientific' events became more frequent.

Mention should be made here of the voyages of Turnefor during 1700–2 from which came descriptions of the plant life in countries surrounding the Mediterranean including characterizations of high altitude regions. 'Cosmographies', descriptions of different countries, which differed little in quality from medieval accounts and which contained alongside reliable facts a good deal of fantasy, enjoyed great popularity during this period.

The greatest achievement of geographic thought during the middle of the seventeenth century was the *Geographia Generalis*, the 'general geography', of a young Dutch scientist, Bernhardt Varen or Varenius (1622–50). This text was published in 1650 and constituted the first attempt to derive a theoretical account of geography and to formulate its subject matter and objectives. According to Varenius, the subject of geography is the earth's surface including dry land, water and atmosphere. Moreover, it should be studied in a broad context taking into consideration the entire earth as well as its individual sections. In this way Varenius aimed to link together general geography and sectional geography (the regional geography of nations); he also outlined a division into physical and social geography. Varenius gave a systematic account, in a general geography context, of the natural phenomena on the earth's surface.

Varenius' ideas were ahead of his time and were not immediately accepted. They were, however, noticed by Newton, who published Varenius' book in 1672 in England, and in 1718 it was translated into Russian by command of Peter the Great. Varenius exerted an undoubted influence on V. N. Tatishchev (1686–1750), who can be considered the first Russian scientific geographer. Like Varenius, Tatishchev divided geography into general and sectional geography and, in addition, distinguished three branches of geography: mathematical, physical, and political geography. Tatishchev was a vigorous proponent of the economic and general administrative importance of geography and devoted himself to the first geographic description of Russia.

The second half of the eighteenth century was characterized by rapid development of capitalism and the rise of productive forces in the major European countries. This was the period of industrial revolution in England and bourgeois revolution in France. The growing interest in the study of natural resources led to the inclusion of naturalists in many expeditions, some of which were devoted to scientific tasks, namely astronomy and geodesy, but also botany and zoology. The so-called academic expeditions equipped and organized by the Russian Academy of Sciences during 1768–74 provide examples of the most comprehensive natural science investigations made during this period. P. S. Pallas, I. A. Gyuldenshtedt, I. I. Lepekhin and other prominent naturalists took part in these expeditions. The accounts of Georg Forster in 1780 and other traveller-naturalists contributed substantially to the knowledge of physical geography.

Towards the end of the eighteenth century geology was established and the first geological maps appeared. The study of the vast amount of material available on flora and fauna enabled a few natural scientists, in particular P. S. Pallas, to come very close to the idea of the evolution of the organic world. In 1755 Immanuel Kant proposed his theory of the origin of the solar system, thereby, in the words of F. Engels, making the first breach in the hardened wall of concepts concerning nature. However, the same Kant, whose outlook was characterized by the simultaneous acceptance of contradictory philosophical principles, contributed to the consolidation of the metaphysical concept of geography in the guise of chorological science, i.e. one which deals only with the spatial evolution of phenomena. According to Kant, history is a description in time; geography is a description in space. Because of his reputation Kant's views affected strongly the subsequent development of geography.

The new achievements in natural science during that period proved insufficient to bring about a breakthrough in geography. The division between its two major directions continued to exist. On the one hand there appeared general geography studies rather diverse in character (physical descriptions, natural histories of the earth, etc.); on the other hand national geographies of various kinds were published. The content of the latter gradually improved, and fantasy gave way to fact. The best examples of the complex 'national-geographic' monographs of the time are *Description of the Land of Kamchatka* by S. P. Krasheninnikov published in 1755 and P. I. Rychkov's *Topography of Orenburg* published in 1762. There was a noticeable trend away from the previous method of geographic description by political units, and attempts were made to construct descriptions according to natural territorial categories. Gatterer, in his *Outline of Geography* published in 1775, used rivers and mountains as natural boundaries, while Russian textbooks on geography, beginning with that of A. F. Bishing in 1766, introduced a schematic zonal division of the country into three regions, northern, central, and southern.

In the main, geographic descriptions and textbooks published during the late eighteenth century represent dry listings of diverse pieces of information and statistical data, among which natural conditions receive negligible attention.

The ideas of M. V. Lomonosov (1711–65), who adhered consistently to a materialist outlook in the study of natural phenomena and developed the idea of universal variability of nature and who was the first to adopt a historical approach to the explanation of contemporary geographic facts, had momentous methodological significance for the geography of the eighteenth century. In his attempt to explain the origin of mountains, minerals and soils, and climatic changes, Lomonosov anticipated trends in natural science. He assisted directly in the development of Russian geography through his management of the Geography Department of the Academy of Sciences and through his cartographic activities. Lomonosov developed a scheme for the settlement of the Arctic and a programme for a comprehensive study of the natural productive forces in Russia.

During the first half of the nineteenth century interest in the unknown areas of the globe grew considerably, due mainly to the search by the large European states for new markets and raw materials, and also to the intensification of colonial expansion. Furthermore, the demand by the capitalist economy for new sources of mineral raw materials and fuels stimulated exploration. It was during this period that the first geographic societies were established in France, Germany, England and Russia. In addition, in 1835 the first geological survey department was established by the British government; similar departments were soon established in other countries as well. The end of the nineteenth century saw the outstanding geographic investigations of Alexander von Humboldt in Central and South America during 1799–1804. These were followed by a large number of other expeditions involving scientific research objectives.

The intensive accumulation of diverse natural science information created a base for theoretical generalization in various branches of geography and gradually led to the differentiation of the original unified science of geography. Von Humboldt's work laid the foundation for the study of climatology and biogeography. Ocean voyages, in particular the Russian circumnavigations of the world, led to the establishment of oceanography. Considerable success was obtained in the study of certain geomorphological processes, although geomorphology was established as a formal discipline very much later. Towards the end of this period zoogeography was established, largely due to the investigations by the prominent Russian naturalist K. F. Rulye (1814–58).

An event of considerable importance in the progress of disintegration of unified geography was the differentiation of economic geography from political geography. Although regional descriptions of nations appeared in textbooks as before, grouping together data about nature, culture, economics and politics, at the beginning of the nineteenth century a distinct separation can be observed in the area of theory and in the analysis of laws. Characteristic of this separation are certain studies in regional geography. In Russia these studies developed in two main directions. One of these, represented by the work of E. F. Zyablovski in 1807, E. F. Kankrin in 1834, and others, was based on the zonal-climatic principle and belongs essentially to physical geography. The other direction in regional studies, economic geography, stems from the monographs published during 1818–48 by K. I. Arsenyev (1789–1865).

The trend towards differentiation in geography was also very strong in tertiary education in Russia. The first approach taken by Russian academic geography was descriptive-statistical, i.e. inclined towards economic geography, and was followed by the history and philology faculties. The second, the physical-geographic approach, began in 1835 and was followed by the physics and mathematics faculties. The Russian physicist and oceanographer E. H. Kh. Lents (1804–65) saw the study of the physical laws governing phenomena on the earth's surface and of 'the accessible deep layers of the earth' as the principal task of physical geography. Earlier still I. A. Dvigubski considered physical geography as a branch of physics. The reason for this situation was the indeterminate content of physical geography, which was often confused with physics and geophysics. There is no doubt, of course, that conditions were ripe at this time for the separation of this branch of natural science from descriptive-statistics, which was oriented to social or cultural geography.

Engels has shown that the differentiation of science, i.e. its separation of nature into individual units, was the principal reason for the dramatic success of contemporary natural science. Engels continued:

> But this method of investigation has also left as a legacy the habit of observing natural objects and natural processes in their isolation, detached from the whole vast interconnection of things; and therefore not in their motion, but in their repose; not as essentially changing, but as fixed constants; not in their life, but in their death. And when, as was the case with Bacon and Locke, this way of looking at things was transferred from natural science to philosophy, it produced the specific narrow-mindedness of the last centuries, the metaphysical mode of thought.[8]

For geography, this process of differentiation of knowledge had yet another specific result: it created a real threat to the very existence of the subject. Geography gradually lost one objective after another; the study of the strata of the earth and the history of their development was transferred to geology, the atmosphere and its processes into meteorology and climatology, the study of the seas into oceanography, etc.

Natural science, however, gradually gave up the metaphysical view of nature which necessarily led natural scientists to the realization of the need to study natural phenomena in mutual associations and as dynamic entities. In 1830 Sir Charles Lyell showed that geological history consists of a successive chain of phenomena, and rejected the catastrophic theory according to which changes in nature occur by a complete catastrophic destruction of earlier forms and of successive 'acts of creation'. In 1842 the mechanical theory of heat appeared, constituting the first proof of the transformation of energy. The discovery of the organic cell and, finally, the emergence of Charles Darwin's theory of evolution in 1859 constituted, according to Engels, the decisive discoveries which shattered the metaphysical concept of natural science. Engels also attributed credit for the dissolution of the metaphysical concept to the various expeditions and voyages after the mid-eighteenth century, and to the development of the comparative method in physical geography, especially the work of von Humboldt.[9] It is well known, of course, that Darwin's theory drew essentially on material which he collected during his round-the-world voyage in 1832–6.

The greatest achievement in the area of geographic generalization up to the mid-nineteenth century belongs to the great German naturalist Alexander von Humboldt (1769–1859). He established the principle of complex investigation of natural phenomena and strove to prove that geography is not merely a sum total of various scientific disciplines but an independent science concerned with the relationships between natural phenomena on the earth. According to von Humboldt, geography must give 'a general account of global phenomena and their mutual interdependence'. He also stressed that the peculiarities of nature in various parts of the earth's surface must be regarded as parts of a whole, i.e. he assumed the unity of general geography and sectional or regional geography. Von Humboldt established the concepts of latitudinal zonality and high-altitude zonality of the plant and the animal world in relation to climate. He was also responsible for magnificent examples of integrated descriptions of steppes, deserts, and tropical forests on a comparative basis.

Von Humboldt's outlook must be viewed not as geographic in the old sense of general geography, but as physical-geographic. Specifically, in his work is found the justification for physical geography as an independent science. Only when he separated the physical-geographic laws from the social was he able to ascend to the level of broad

physical-geographic generalizations. At the same time the establishment of physical geography as a separate science did not prevent
von Humboldt from working in narrower areas of geography, i.e.
climatology and biogeography. Despite von Humboldt's work, the
prevailing outlook in the first half of the nineteenth century was still
metaphysical, and the growing differentiation of geography, together
with an insufficient development of its new branches, proved inadequate for the construction of a basis on which the process of
geographic synthesis could operate. These are reasons why von
Humboldt had virtually no followers in his day and the ideas of
another prominent German geographer, Karl Ritter (1779–1859), the
author of several classical monographs concerning Asia and Africa,
attained much wider popularity.

The outstanding geologist and geographer I. V. Mushketov, in his
evaluation of the significance of von Humboldt and Ritter for
geography, wrote in 1886:

> The vast difference between them resides in the fact that von Humboldt
> based his conclusions on personal observations of nature and made use
> of a large selection of sources. Being the foremost expert of his age in
> the various branches of knowledge, especially of physical geography
> and geology, he knew how to evaluate these sources in a highly critical
> manner. In this he was completely opposite to Ritter, who virtually all
> his life discoursed about nature, viewing it through the windows of his
> study. Von Humboldt, in addition to his brilliant generalizations,
> supplied science with a mass of new facts. Ritter merely classified the
> existing data, illuminating these with his own ideas. Von Humboldt
> studied the most diverse phenomena in nature from the point of view
> of their interdependence and tried to explain their origin; Ritter based
> everything on form, on external configuration, without attempting to
> explain homologous forms. Von Humboldt recognized the effect of
> man on nature, but did not try to create an independent science out of
> this realization ... Ritter, on the other hand, wanted to establish a new,
> implausible science on this basis.[10]

Ritter did a good deal to popularize geography and tried to prove
its autonomy, but his concepts carry the imprint of an idealistic
philosophy. Ritter's views on geography are permeated with a
teleological flavour, i.e. they assume the existence of an initial concept,
a god-given predetermination of natural and social phenomena. The
earth is for him 'the abode of man', and natural phenomena are
investigated only in so far as they can be used to explain differences
in the life, culture and economic activity of man. These ideas assisted
the development of the concept of geography as a 'unified' science, i.e.
uniting nature and man, and also as a 'chorological' science, i.e. one
concerned with spatial relationships. But these sciences could not

provide a foundation for the exhaustive investigation of the internal relationships of geocomplexes.

During the second half of the nineteenth century the crisis in geography, caused by its fragmentation, worsened. At the same time geography moved through a crucial period of change. The old descriptive-chorological geography entered into conflict with the idea of the mutual interdependence of phenomena and the ideas concerning evolution, especially after the publication of Darwin's *The Origin of Species*. In 1856 P. P. Semyonov-Tyan-Shanski admitted that geography was not a unified science but a group of sometimes loosely related yet independent sciences including natural geography (astronomical geography and physical geography) and social geography (ethnography and 'statistics'). Even physical geography did not really represent something integral and defined; and under this name were combined a fairly large group of independent branch disciplines. The development of these was stimulated by the requirements of a vigorously growing capitalist economy and new types of transport and communications.

During this period, central meteorological services as well as geological exploration services were established in many countries. Systematic investigation of water supplies and river systems began in the highly developed capitalist countries; soil survey, etc. were also initiated. Towards the end of the nineteenth century most European countries possessed large-scale topographic maps. The importance of these maps for geography grew enormously when the cross-hatch representation of relief was replaced by contour lines.

Among the individual physical-geographic disciplines climatology was particularly successful; regional monographs concerning climate appeared, among them K. S. Vyesolovski's *The Russian Climate* published in 1857. Furthermore the processes of atmospheric circulation and climatic variations were studied and in 1884 the outstanding work of A. I. Voyeykov, *The Climates of the Earth, and Russia in Particular*, appeared. It exerted an enormous influence not only on the development of climatology but also on the progress of hydrology and physical geography.

The efforts of many scientists both foreign (especially W. M. Davis) and Russian (including P. N. Kropotkin, V. V. Dokuchayev, I. V. Mushketov and A. P. Paulov) led to the establishment of geomorphology. In 1885 S. N. Nikitin produced the first geomorphological regionalization of European Russia, while in 1889 A. A. Tillo published the first hypsometric chart of this territory.

Scientific soil study, the foundation of which was laid by V. V. Dokuchayev, commenced in the 1880s. During this period biogeography developed vigorously. In zoogeography special mention should be made of the work of N. A. Severtsov (1827–85). Geobotany, the study of plant societies, was formulated during the 1860s with the work of F. I. Ruprecht and I. G. Borshchov, but its vigorous growth is associated with the complex investigations of the steppe directed by V. V. Dokuchayev and carried out during the 1880s and 1890s by his students A. N. Krasnov, G. I. Tanfilyov, S. I. Korzhinski and I. K. Pachoski.

Unprecedented interest in geography developed towards the end of the nineteenth century, characterized by intense discussions about its subject matter and scope. Many diverse concepts emerged of which there are two main schools of thought.

Some western European geographers, in particular F. Ratzel, sought to preserve the unity of geography on the basis of idealistic principles, reconstructing it on the teleological concept of Ritter and on the ideas of geographic determinism (a theory concerning the decisive influence of natural conditions on social phenomena). Ratzel attempted to establish that 'impossible science' which brought such opposition from I. V. Mushketov. The supporters of this position defined the object of geography not from its inherent characteristics but from its relationship to man including the exclusion from geographic studies of all natural objects which at the given moment were of no direct interest to man. Such a definition is obviously not in keeping with the character of natural science but anthropocentric, and leads to the disappearance of physical geography as an independent science.

Other geographers sought the panacea for geography in an ancient chorological concept, i.e. they defined geography as a science concerned with the content of terrestrial space including the spatial relationships between diverse objects and phenomena, regardless of their nature. According to this position, geography is likewise a unified science, but only within the framework of the regional geography of nations. Regional geographies of nations can, of course, group together diverse facts about nature and about socio-economic phenomena. But in the study of governing laws, natural and social phenomena need to be separated, since they develop in accordance with different laws,* disregarding those general laws of matter formulated by the Marxist philosophy. For this reason the supporters of the chorological concept were forced to deny the need for the study of

---

* This important concept must be appreciated to understand fully trends in modern Soviet geography. It has recently been discussed by V. A. Anunchin in his doctoral dissertation, by his examiners, and by other interested geographers, (Saushkin, 1963; Report on discussion of Anunchin's book, 1961)

the laws of general physical geography so as to avoid the complete subordination of social phenomena. In such a system logic demands that general physical geography should thus be rejected, and a unified geography reduced to the regional geography of nations. The most ardent supporter of this position was the German geographer A. Hettner.

More progressive views were expressed by only a few foreign geographers, and for the most part these views remained undeveloped. F. von Richthofen (1833–1905) wrote in 1883 that the subject matter of geography is the surface of the earth including the lithosphere, hydrosphere and atmosphere, together with living organisms. Furthermore he indicated that the task of geography involved the explanation of the mutual relationships between these components, both along general physical-geographic and regional lines. E. Reclus (1830–1905) also stressed the exceptional importance of studying natural processes and their mutual relationships, but thought that the subject of geography must be the entire globe. There were others, including Gerland, who also insisted that geography must concern itself with the globe as a whole. The differentiation of such a large number of independent sciences concerned with the earth (e.g. geodesy, geophysics, geology, etc.) is clear proof of the untenable nature of this view.

The contemporary direction in geography originated in Russia. This direction was sharply opposed to the official, academic geography which in the universities (Moscow from 1757 and St Petersburg from 1819) was treated as a service discipline with only a few students taking courses. During the 1830s and 1840s, moreover, the teaching of geography was virtually abandoned. New departments of geography were not created until after 1884, first at Moscow University and then at St Petersburg University; even later, departments were established in the universities at Kharkov, Kazan, and a few others. These events reintroduced discussion about the subject matter and tasks of geography.

In 1887 a special committee of inquiry established under the chairmanship of P. P. Semyonov-Tyan-Shanski set out to decide in which faculty geography should be placed. The committee noted, in particular, that geography is concerned 'not with isolated phenomena but groups or associations of phenomena, as well as with the laws of interaction among such phenomena'. It concluded that geography should be included in the physics and mathematics faculties. By its decision the committee accepted the view that geography is a *natural science*. This alone, however, could not solve the problems confronting geographers including the diversity of opinions which continued to be expressed as before.

E. Yu. Petri (1854–99), appointed in 1887 to the Chair of Geography and Anthropology at the University of St Petersburg, held the view that 'the task of geography is to comprehend the sense and the life of the earth', i.e. he sought to re-establish the concept of geography as a universal science about the earth. A similar view was expressed in 1892 by the father of the Moscow University's School of Geography, D. N. Anuchin (1843–1923), even though he later narrowed down the scope of the subject. By contrast, in 1896 E. I. Chizhov viewed geography as the regional geography of nations in the old 'chorological' sense.

The new direction in geography in Russia evolved outside the university mainstream; it was born of the experience gained through regional investigations of a natural science kind in the course of solving practical problems; it rested on the materialist traditions of Russian science. Its followers were naturalists, for the most part bio-geographers and soil scientists, who did not really consider themselves geographers. As early as the 1840s and 1850s many zoogeographic and geobotanic investigations were given a broad physical-geographic character in terms of deep analysis of the relationships between plant and animal life and the components of the environment. The study of these relationships led some scientists to distinguish types or kinds of localities and to regard them as territorial combinations of physical-geographic components to which certain groups of plants and animals are associated. These ideas exhibited the concept of the geocomplex.

One of the first studies of this kind was a monograph by E. A. Eversman (1794–1860) entitled *The Natural History of the Orenburg District* and published in 1840, which included a synthesis of natural conditions in terms of zonality,* and distinguished within each zone various types of natural complexes (e.g. clayey, alkaline steppe, sandy steppe, etc.). The remarkable work by N. A. Severtsov, *Periodic Phenomena in the Life of Animals, Birds and Reptiles in the Voronezh Region*, in which various types of localities are specifically described, was published in 1855. I. G. Borshchov developed during 1860–5 a comprehensive description of the natural complexes in the Aral-Caspian region. In 1871 M. N. Bogdanov divided the chernozem area of the Volga country into 'locality types'. The significance of such types in geographic investigations was discussed by P. P. Semyonov-Tyan-Shanski. P. N. Kropotkin insisted in 1893 that in teaching

---

* Zonality is the variation in physical-geographic conditions in a north–south direction. It is discussed in detail in chapter 2.

geography, attention should be concentrated on the review and analysis of 'landscape types'; however he never elaborated on the meaning of the idea.

The work of Russian geographers during the latter part of the nineteenth century preceded the appearance of Dokuchayev's teaching concerning the zones of nature. In 1877, N. A. Severtsov published a zonal zoogeographic schema which, in the author's own words, was based on physical-geographic principles. Furthermore, P. P. Semyonov-Tyan-Shanski and N. A. Severtsov investigated and described the high-altitude zones in the mountains of central Asia. These geographers must be considered direct forerunners of V. V. Dokuchayev whose work laid the foundation for the establishment of contemporary geography.

The geographic ideas of V. V. Dokuchayev and the emergence of landscape science

V. V. Dokuchayev left behind him an exceptionally large and diverse legacy of scientific investigation, but for the subsequent development of geography his discovery of soil as a separate natural body and his establishment of the principles of the scientific study of soils had a profound influence. During the late 1870s, before his work appeared, soil was regarded either as the eroded surface layer of bedrock or as a tillable layer. Even von Humboldt, as well as many other foreign geographers after him, did not distinguish soil from bedrock and did not treat it as a separate geographic component. Soil, which according to Dokuchayev is a function of all the other geographic factors—bedrock, climate, the organic world, etc.—constitutes the clearest expression of the geocomplex; it is its product and its mirror. Without soils the concept of the geocomplex can be neither meaningful nor complete.

Dokuchayev's grasp of the concept of the geocomplex was entirely natural, precisely because soils constituted for him a starting point. As a spontaneous materialist and dialectician, Dokuchayev could not fail to observe the 'negative' results of the then well-advanced fragmentation of science. He wrote that the science of the nineteenth century, having achieved enormous success in the study of individual bodies and phenomena, lost from view that genetic bond and those inter-relationships between the bodies and phenomena of nature, the study of which constitutes the 'essential knowledge of nature', 'the best and highest beauty of natural science'.[11]

Dokuchayev came out against the metaphysical fragmentation in the study of natural phenomena not only as a theoretician, but also as a practical scientist. His entire scientific activity was inseparably linked with the problems of agriculture, and his ideas developed during his investigations of the Russian steppes; these investigations were aimed at drought-control methods and the establishment of a scientific foundation for soil science. Despite this, Dokuchayev's approach to nature and its resources was far from utilitarian. Even in his early work in the 1870s he pursued the idea that 'to study the given phenomena or the given object of nature with merely a utilitarian purpose has always been and always will be a most serious error, since both phenomena and objects exist in nature quite independently of us.'[12]

To learn to control natural processes, Dokuchayev said, one must 'honour and study the one, whole and indivisible nature and not its fragments.'[13] On these principles, he developed a geographic basis for the rational management of agriculture in the different natural regions of the country.

Dokuchayev was the first to realize in practice the complex principles guiding the study of natural conditions in different areas: first in his investigations of the Nizhnegorod region during 1882–6, later, during 1888–94, in the Poltava province, and later still, during 1892–8, by organizing a special steppe expedition, which involved carrying out complex investigations of the steppes and employed various methods of field investigation. These expeditions were a veritable school for research geographers of a new breed. During this period there evolved the Dokuchayevian school of geography, the members of which included the leading geographers of the day. Among them were Dokuchayev's students, N. M. Sibirtsev (1860–1900), A. N. Krasnov (1862–1914), G. N. Vysotski (1865–1940), G. I. Tanfilyov (1857–1928), G. F. Morozov (1867–1920), V. I. Vyernadski (1863–1945) and K. D. Glinka (1867–1927). In addition, there were L. S. Berg (1876–1950), S. S. Neustruyev (1874–1928), B. B. Polynov (1877–1952) and other prominent geographers.

Dokuchayev and his successors interpreted and creatively developed valuable contributions to geographic thought. At the very end of his creative period Dokuchayev realized the necessity for an independent science concerned with the relationships between the living and non-living components of nature and the laws of its genesis. He intended to devote much of his attention to this task, but was unable to do so. The initial part of this work was the investigation of the zones of nature, elaborated by Dokuchayev in a series of papers appearing between 1898 and 1900. Today these studies are so well known that there is no point in discussing them in detail. It is only necessary to mention that Dokuchayev first treated zonality as a natural law: every natural or natural history zone constitutes a regular natural complex in which the living and non-living aspects of nature are closely associated one with another. Such views regarding geographic zones were not expressed by Dokuchayev's forerunners, even those of von Humboldt and later Severtsov came fairly close to Dokuchayev's concept.

Dokuchayev himself did not give the new science its name, but stressed that it must by no means be confused with 'geography, which is developing in all directions at once'. This, at first glance, paradoxical statement becomes quite comprehensible if it is kept in mind that the official academic geography was then passing through a crisis and had no real prospects for future development.

Many of Dokuchayev's students and followers sought to prove that his ideas constituted the essence of geography. In 1903, the year of Dokuchayev's death, A. A. Yarilov wrote of Dokuchayev's influence on his contemporaries and their subsequent work in geography. Moreover, L. S. Berg showed that Dokuchayev's natural history zones are landscape zones, and that the new science concerned with their relationships—the science which Dokuchayev tried to establish—is landscape geography. Elsewhere, L. S. Berg names Dokuchayev as the founder of contemporary geography.

Of great interest are the remarks of one of Dokuchayev's closest followers, A. N. Krasnov, who in 1889 was appointed to the Chair of Geography at Kharkov University. By contrast with his academic colleagues, Krasnov did not attempt to champion the old geography which 'ran in all directions at once'. His opinion was that the old geography represented a 'collection of diverse data' about various countries, 'beginning with the direction of prevailing winds and ending with the description of the implements used in cutting roast beef.' In 1889 Krasnov concluded that this old geography had in fact disintegrated into a multitude of independent sciences, and that the recent achievements of natural science and technology must be used to create a new soil science; its task would be the investigation of the causal and genetic relationships between terrestrial phenomena. In his view soil science should be divided into general and specific (or regional) studies.

These statements by Krasnov do not give an indication of the significance of Dokuchayev's ideas for the development of modern geography. The influence of these ideas on Krasnov, however, dating back to his student years when he participated in Dokuchayev's expeditions, is undoubted. In 1895 Krasnov stated that geography must study the territorial combinations of natural phenomena, i.e. geocomplexes; he included the steppes, deserts, etc., among such complexes. Later, in 1910, Krasnov gave a brief description of the major geocomplexes, which more or less correspond to Dokuchayev's zones.

G. F. Morozov, one of the outstanding members of Dokuchayev's school of geography, wrote of Dokuchayev at the beginning of this century that contemporary geography, with its tendency to study the interrelationships between diverse objects and phenomena in nature, found a most brilliant spokesman in this great founder of soil science. In his writings on geography published between 1909 and 1920, Morozov stressed particularly its genetic character and its real, practical significance. As far as can be ascertained, it was Morozov who originated the idea of *applied geography*; he included in it forestry and land reclamation.

The endeavour and the ability to combine the development of theory with practice are the characteristic features of the Dokuchayevian school. Against the background of the study of the geocomplex, Dokuchayev and his followers developed scientific foundations for the evaluation of agricultural areas including drought control and the rational planning of land use, and the theory of forest and steppe-forest management.

The Dokuchayevian principle of zonality proved particularly valuable in the early 1900s. It was used extensively in the solutions to various agricultural and forestry management problems; during the pre-October period it provided a basis for the development of physical-geographic regionalization. Further research into the zones of nature was accomplished during Dokuchayev's own lifetime; this includes work by N. M. Sibirtsev, G. I. Tanfilyov and G. N. Vysotski, and later by L. S. Berg, A. N. Krasnov and others.

The study of landscape, landscape science, was the next logical step in the development of the Dokuchayevian concept of the geocomplex and particularly the zones of nature. It is pertinent that between 1904 and 1914 the scientific concept of landscape was formulated almost simultaneously, yet independently, by a number of geographers: G. N. Vysotski, G. F. Morozov, L. S. Berg, A. A. Borzov and R. I. Abolin. To a Dokuchayevian geographer it necessarily became obvious that the surface of the earth consists of objectively existing territorial units, each of which constitutes a regular and specific combination of objects and phenomena of nature. While these ideas were obvious to Dokuchayev, he did not, however, make explicit his views about geocomplexes and did not formulate their definition.

For the introduction of the concept of 'landscape' into science we are indebted mainly to L. S. Berg who as early as 1913 clearly expressed the view that the subject of geography is the study of landscape. Berg defined the natural landscape at that time as an area in which the character of relief, climate, plant and soil cover merges into a single harmonized entity reappearing regularly over a defined zone on the earth's surface. In addition, Berg emphasized the association between landscapes and natural zones; he defined the latter as 'regions in which a specific landscape predominates' and gave them the name *landscape zones*.[14]

G. F. Morozov founded modern forestry on the basis of principles of landscape geography. Morozov was the first scientist to define forests as 'geographic phenomena' and to differentiate them into types of geocomplexes. He considered that the ultimate result of the study of the natural history of a territory is its subdivision 'into a set of *landscapes* or *geographic individuals*'. These ideas exerted a strong influence on the development of Berg's views; Morozov, in turn, found in Berg's work theoretical support for his own theories concerning forestry.

G. N. Vysotski, well known for his work in steppe-forest cultivation, soil science, geobotany, hydrology, etc., shared Morozov's ideas. As early as 1904 Vysotski developed independently the concept of landscape which he designated by the Russian term 'myestnost' or 'okrug'. According to Vysotski myestnost differ from one another in terms of the character of the groupings of myestnost-inception types; myestnost-inception types correspond to the modern concept of morphological units of landscape which will be discussed in a later section. Vysotski was the first to propose the construction of maps of myestnost-inception types, or phyto-topological maps—landscape maps in the modern sense. He thought that such maps constitute the scientific basis for the organization of agriculture.

Further developments of the pre-October period affecting the emergence of landscape science will not be considered. The study of landscape may be regarded as a new stage in the development of specific or regional geography. But development in geography was not limited to regional studies.

In 1910 P. I. Brounov (1852–1927) wrote that the subject matter of geography is the 'external terrestrial envelope' which consists of the lithosphere, hydrosphere, biosphere and atmosphere, all interpenetrating and continuously, mutually interacting. The study of these interactions is, according to Borunov, one of the fundamental directions of physical geography.[15]

Borunov's interesting ideas did not find wide acceptance at the time, yet a substantial rift occurred between landscape geography, which had achieved great prominence, and general geography. L. S. Berg considered that landscape science constitutes geography; general geography in his opinion was not a geographic discipline at all. This view exemplifies the narrowness of Berg's ideas. It is true that some geographers, in particular D. N. Anuchin, A. A. Borzov and A. N. Krasnov, stressed the need for investigations into the laws of both general and regional physical geography. None of them, however, succeeded in achieving a practical synthesis of these two directions.

In 1914 R. I. Abolin came very close to an accurate understanding of the relationship between the laws of general and regional physical geography. According to Abolin, landforms, soils, vegetation and other components interpenetrate so closely and so strongly affect one another that they jointly comprise 'a single complex geographic phenomenon, an intricate complex formation, an epigeneme covering the entire land mass between the equator and the poles.'[16] The 'epigeneme', i.e. the complex landscape envelope, divides according to latitudinal zonality into 'epizones' and, according to geologic history, into separate 'epiregions'. In every region, owing to the variation in local conditions, different 'epitypes' can be observed. Finally, the indivisible, homogeneous segment of the terrestrial surface is designated by Abolin as the 'epimorph'.

Abolin thus created a consistent system of geographic classification of the earth's surface from the landscape envelope to an elementary geocomplex, 'facia' in modern terminology. He employed these ideas in developing a genetic-based landscape classification of marshlands. His system, however, was not applied to a variety of terrain conditions and for this reason geographers were generally unaware of its existence.

Despite the fact that the relationship between general and regional geography was not entirely resolved in pre-October Russian geography, the basis for a materialistic concept of physical geography developed during this period. The idea of 'unified' geography, i.e. mixing physical and cultural studies, which nearly always had been associated with the idealist outlook, was not popular among Dokuchayevian geographers, although individual scientists including L. S. Berg were somewhat inconsistent in this regard. A distinct attitude to this problem emerged in the discussions and research concerning regionalization during 1908–10. G. N. Vysotski insisted that physical-geographic and economic regionalization should be carried out separately: it is wrong to create a 'mixture' of the two, since each arises from a different system of laws. This view was shared by P. V. Otoski, S. S. Neustruyev, G. F. Morozov, L. I. Prasolov and others.

Summarizing the achievements of the pre-October or Dokuchayevian period in the development of scientific geography, we find that the establishment of contemporary physical geography has been accomplished on the basis of the theory of landscape. The methodological significance of this theory resides in its concern with the idea of a natural complex and the idea of mutual interaction of processes. This theory assisted in overcoming metaphysical and idealistic influences and consolidating the dialectical-materialist outlook both in physical geography and in associated disciplines of natural science. The concept of landscape, even during the pre-October period, began to gain acceptance among natural scientists, but landscape science was yet to come.

The development of landscape science in the U.S.S.R.

After 1917 geography in the U.S.S.R. entered a new phase. The investigation of productive forces became a national task and assumed a planning character. During the 1920s expeditions were organized to study principally the little-known marginal regions of Russia and the autonomous republics.

In old Russia, geography did not exist as a specialized discipline; nearly all geographers were self-taught, and for many geography was only an incidental occupation. In 1918 the Soviet government created in Leningrad the Institute of Geography, the first institution of higher education in geography. In 1925 it merged with the Leningrad State University being accorded the status of a faculty. Faculties of geography were also established in many other universities. The U.S.S.R. was also the first country to establish geographic research institutes within the system of the U.S.S.R. academies of science in the Union Republics and in the universities. This progress, together with the gradual reconstruction of geographic theory on the foundation of dialectical materialism, contributed greatly to the subsequent development of landscape science.

The first historical period in Soviet landscape science, including mainly the work of the 1920s, did not initially provide new achievements in theory although the concept of landscape began to enter into field work. Field investigations carried out by specialists in various disciplines confirmed the need for the differentiation and detailed study of geocomplexes and enriched the concept of landscape science. The most important result of the many field studies associated primarily with the development of new farmlands, land reclamation, and other practical tasks, was the appearance of the first landscape maps. The initiative for the construction of these came from B. B. Polynov, I. V. Larin and R. I. Abolin. These large- and medium-scale maps were produced for the most part by the method of field landscape survey, i.e. by on-the-spot investigation of selected terrain sections which differed with regard to relief, bedrock, soil or vegetation.*

It must be stressed that the first landscape maps did not result from specialized landscape studies but emerged spontaneously owing to practical need. B. B. Polynov insisted that practical developments such as land reclamation should be based on landscape maps. Every boundary on such maps conveyed a large set of mutually interconnected indices of natural conditions; the landscape map thus constituted a synthesis of these conditions. Moreover, landscape mapmaking during this period was carried out by various geographers without any attempts at co-ordination, and as a result the maps lack uniformity.

Progress in landscape surveying revealed a great diversity of geocomplexes and generated the need to establish a hierarchy of landscape units. As a result, in large-scale mapping the object of study is the most elementary geocomplex, which B. B. Polynov and I. M. Krasheninnikov designated as 'elementary landscape' and which I. V. Larin called the 'micro-landscape'; the latter corresponds to R. I. Abolin's 'epimorph'.†

Centralized economic planning, the establishment of autonomous republics, and the reconstruction of the administrative system in the country, greatly stimulated physical-geographic regionalization. Even during the first decade of Soviet power there appeared numerous systems for the regionalization of individual republics, regions and provinces. It is true that these systems for the most part have an empirical nature and reflect no general theoretical principle. One of the best examples of this work is the physical-geographic regionalization of the Orenburg region carried out by S. S. Neustruyev in 1918 and based on L. S. Berg's ideas of the geographic landscape. In the introduction to this work Neustruyev gave the first popular exposition of the concept of landscape.

Most pertinent to the development of the theory of physical-geographic regionalization and to the study of the laws of physical-geographic terrain differentiation was the introduction of the so-called

---

* Similar procedures were used by the Michigan Land Economic Survey in the 1920s and early 1930s. (Davis, 1969, pp. 20-1) The Soil Conservation Authority of Victoria, Australia and the CSIRO Division of Land Research employ extensive field methods in preparing their reports and maps of land systems. (Gibbons and Downes, 1964, pp. 17-20; Christian and Stewart, 1968, pp. 260-3)

† These units correspond to the familiar land element or land component. An excellent, brief review of the units recognized by various Western survey organizations has been prepared by the Oxford–MEXE group. (Brink *et al.*, 1966, p. 316)

principle of *azonality* (also called provinciality). This principle relates to longitudinal terrain differentiation and is associated with climatic and tectonic-geomorphologic variation in the landscape envelope. The introduction of azonality made it possible to supplement and improve the accuracy of systems for the natural regionalization of Russia which previously were almost exclusively based on the zonality principle. Geographers began to devote attention not only to the zonal but also to the azonal or provincial factors of landscape formation.

The field investigations of the 1930s and the early 1940s up to World War II include novel and successful applications of the land-scape method in the solution of important problems associated with the national economy. During the 1930s, surveys of food producing areas were implemented. These led to the accumulation of a large volume of factual data which allowed L. G. Ramyenski to formulate several new developments in landscape theory. The principles of landscape science were applied to the investigation of sandy areas in desert-steppe regions, in the study of tundra, etc. There was still no significant progress, however, in the development of methods in land-scape research.

The 1930s and early 1940s were also characterized by increased interest among geographers in the methodological and theoretical aspects of landscape science, an interest which gave rise to intense methodological discussions. This interest derived from the appearance during the 1930s of a new generation of highly qualified Soviet geographers. They interpreted the subject matter and the tasks of geography from the dialectical-materialistic position and were opposed to the legacy of classical science (metaphysical ideas, etc.). This trend was not without 'extremist' views, as exemplified by the groundless rejection of the entire work of L. S. Berg owing to a few erroneous interpretations. The major contribution of the 1930s stemming from the discussions concerning the nature of geography was the censure of the chorological concept.

The first volume of L. S. Berg's well-known book, *The Landscape-Geographic Zones of the U.S.S.R.*, was published in 1931. In the introduction to this text Berg refined and augmented his previous definition of landscape. In addition he provided examples of landscape and briefly reviewed the process of interaction between landscape and its components; he touched on the question of landscape development and discussed the evolution of landscape science emphasizing that it had sprung from Russian soil science, particularly the work of V. V. Dokuchayev. In his book Berg takes a very broad view of landscape. Among the examples, he includes both the typological aggregates of elementary geocomplexes appearing regularly in a given zone, e.g. swamps, spruce stands, sand dunes, etc., and intricate, unique geocomplexes, e.g. the central-Siberian plateau and the Valdai Hills.

For a long time Berg's text served as the basis for the subsequent development of the theory of landscape and the duality in his concept of landscape was reflected in the ensuing literature. Some geographers assigned the decisive role in landscape research to the typological approach and used the term 'landscape' in the typological sense. In other words they held the view that landscape is a generic concept and that a landscape is a type of territory which groups together all the similar and homogeneous sections of the earth's surface regardless of their location. According to A. N. Ponomaryov, for example, all the steppe sinkholes with birch and aspen stands constitute a single landscape type. This view was shared by M. A. Pyervukhin who in 1932 published a long methodological paper in which he attempted to examine the problems of landscape science from the Marxist viewpoint.

Parallel with the above interpretation there emerged a different view of landscape which may be called regional. L. G. Ramyenski was the first geographer to develop this view in detail. He demonstrated that a landscape is a relatively complex territorial system consisting of diverse yet mutually consistent, interrelated elementary natural complexes which develop as a single entity. These complexes were called 'epifacies'; they are created on the earth's surface at various localities, i.e. on those elements of relief which have a uniform grade, exposure and configuration of slope. Epifacies are also characterized by uniform biocenosis.* Epifacies correspond in definition to the earlier ideas of epimorph, micro-landscape, and elementary landscape. According to Ramyenski the following common features may be found in every landscape: (1) a regular and uniform division of a surface into localities and their corresponding epifacies; (2) a common origin and development of the entire complex of epifacies; (3) a complex mutual association and continuous interaction among epifacies; (4) a basic character common to the entire complex. Ramyenski emphasized that 'to be fully developed, a landscape requires a territory large enough to contain all the localities characteristic of that landscape in their typical construction.'[17]

---

* 'Biocenosis', often spelled 'biocoenosis', commonly refers to 'the association of plants and animals, especially with regard to a particular area and uniform set of life conditions'. Isachenko also uses the terms *biogeocenosis*, which presumably refers to the same association as above but places more emphasis on the physical base of the 'life community', *phytocenosis*, which refers to relationships among plants, and *zoocenosis*, which refers to relationships among animals.

The term 'urochishcha' was introduced into science by Ramyenski in order to designate intermediate territorial units, i.e. the aggregates of facies. These were associated with individual landforms, e.g. different parts of a large floodplain, etc. Thus Ramyenski derived the principles of *landscape morphology*. His concept of landscape required a geographer to devote himself mainly to the dynamics of landscape, the interaction among the components, and therefore to the processes affecting the distribution of heat, moisture, mineral and organic substances.

The concept of landscape as an integral and unique territory, with a common origin even though morphologically heterogeneous, was developed by S. V. Kalyesnik.[18] At about the same time A. A. Grigoryev proposed the idea of the geographic envelope; he defined the principal features of its structure, thus drawing attention once again to aspects of general physical geography.[19] Employing and developing Grigoryev's ideas, Kalyesnik showed that every landscape, characterized as it is by its unique features, is inseparably bound in all its relationships with the total geographic envelope. Kalyesnik emphasized that landscapes must be directly distinguishable in the field by means of 'landscape surveys', and that landscape scientists are fully qualified to carry out such work. This recommendation was highly significant, since only fifteen to twenty years before, the view had prevailed that the delineation of a landscape constituted a complex task involving the expertise of many different specialists.*

The post-war period of development in Soviet landscape science is characterized both by the increased number of field landscape surveys and by the intensification of theoretical research. From 1945 onwards landscape scientists at the Moscow State University, under the leadership of N. A. Solntsev, undertook detailed experimental landscape survey work; a few years later they began systematic field mapping of landscapes in the Moscow and Ryazan provinces. At approximately the same time, during 1951–2, landscape mapping began at the

* Scientists in Western countries carrying out surveys similar to the Soviet landscape surveys generally work alone or in pairs. For example, the surveys of the Soil Conservation Authority of Victoria or those of the Directorate of Overseas Surveys, England, are accomplished usually by one or two principal investigators. The CSIRO Division of Land Research adopts a team or *integrated* approach, however; team effort throughout the entire project is claimed as the backbone of CSIRO's general-purpose survey method. 'As information about individual resource factors in isolation is of little value, and balanced information about all resource factors in each sub-division of the landscape is necessary, there is need for teamwork, involving a number of specialists, and integration of thought and effort at all stages'. (Christian and Stewart, 1968, pp. 239-40, 260-3)

Leningrad State University. Furthermore, similar work was also undertaken by the geographers of a number of other universities including
those of Latvia, Voronezh and Lvov.

The development of field landscape surveys had a decisive influence
on the subsequent emergence of the theory of landscape. In 1947
N. A. Solntsev published the first theoretical treatment of the results
of fieldwork carried out by Moscow geographers. He further developed
the concept of landscape outlined in the work of Ramyenski and
Kalyesnik; he gave it a new and more precise definition, and considered in detail the questions of landscape morphology.[20] In the
following years the experience gained from landscape investigations
including mapping in the field reflected in full Solntsev's view that
every landscape consists of regularly associated morphological units:
urochishcha and facies. These, in turn, are the objectives of detailed
landscape surveys. During 1947–8 there appeared in the literature
numerous studies devoted to the theoretical problems of landscape
science.

Closely related to landscape science is the study of 'biogeocenosis',
which was developed during World War II by V. N. Sukachev. The
idea itself is not new; it describes the simplest geographical unit and
corresponds to an elementary landscape or facia. The novelty in
Sukachev's approach to the study of these units lies in the attention he
gave to the investigation of the exchange of matter and energy among
the members (components) of the biogeocenosis, and to the role of
organisms in its dynamics. Under Sukachev's leadership a number of
field stations have been established where complex biogeocenotic
studies are carried out.

The foundations for a new branch of landscape science, landscape
geochemistry, were laid by B. B. Polynov.* The study of the geochemistry of landscape, which deals with the migration of chemical
elements in landscape, gives us a new and powerful method by which
we can investigate the interaction among the components of a landscape and its morphological units.

Especially vigorous progress in the development of landscape
science dates from 1955 when the U.S.S.R. Geographical Society
organized in Leningrad the first All-Union congress devoted to the
problems of landscape science. The congress stimulated the development of landscape science in the U.S.S.R. and initiated co-ordinated
investigation of the most important questions facing landscape
scientists. Since that time there have been regular All-Union congresses, at Lvov in 1956, at Tbilisi in 1958, at Riga in 1959, at

* See also Perelman (1961).

Moscow in 1961 and at Alma-Ata in 1963. These congresses provided a forum for the exchange of experience, the discussion of topical problems, the definition of tasks, and the adoption of resolutions in the areas of theory, method of investigation and practical application of landscape science. Problems of general theory and methods in landscape science were also the subject of a special symposium during the Third Congress of the U.S.S.R. Geographical Society at Kiev in 1960.

Today the universities are the principal centres in which landscape science is developing. Universities carry out field investigations, they compile landscape maps and are involved in the physical-geographic regionalization of the country. In addition, they perform tasks in the service of the national economy and publish scientific papers, reports and monographs in various areas of landscape science.

A major task facing Soviet landscape scientists is the necessary intensification of landscape mapping leading to the production of maps at various scales. The resolution of the Third Congress established a committee on landscape mapping whose responsibility it is to co-ordinate work in this area. As its first objective the committee set itself the development of criteria and methods for the compilation of a landscape map of the U.S.S.R. at the scale of 1:4 000 000. This map, which constitutes a vast collective effort and which consolidates the achievements of Soviet landscape science, was presented to the Fourth Congress of the U.S.S.R. Geographical Society in 1964.[21]

A brief account of
the development of landscape science outside the U.S.S.R.

Among foreign geographers, with the exception of a few countries, the concept of the geocomplex is not yet widespread, although a few geographers adopted it by the beginning of the twentieth century. In 1905 the English geographer, Herbertson, wrote that the time was ripe to abandon the study of the geographic distribution of unrelated phenomena and to undertake the study of the distribution of the complexes of such phenomena. The main task of geography in Herbertson's view was the differentiation and classification of similar complexes. He was the first to attempt an outline of the fundamental *types* of natural regions. He took into account common differences in relief, climate and vegetation.*

Foreign geography, however, was dominated by the chorological ideas of A. Hettner, and by other often transparently idealistic con-

* (Herbertson, 1905, 1912, 1913a, b, 1915-16)

cepts. The latter affected the foreign studies in landscape science
which began to gain a measure of popularity during the 1920s and
1930s, mainly in Germany.*

One of the first theoreticians of German landscape science was S.
Passarge (1866–1958) who published a number of theoretical studies
as well as a description of the earth's landscape zones. Passarge treated
zones as the largest units of landscape; he divided them into landscape
regions, and the latter, in turn, into landscapes. Landscapes, according
to Passarge, consist of simpler geographic units—constituent landscapes (*Teillandschaften*) and landscape parts (*Landschaftsteile*)—
which correspond approximately to the urochishcha and the facies of
Soviet geographers. Thus while attaching great significance to the
internal structure of the landscape, i.e. its division into constituent
landscapes and landscape parts, he maintained that their number, size
and mutual distribution determines the economic and even the political
development of the country. At the same time he felt that only those
small fragments at the core of the landscape can be firmly established,
while the method of investigating landscapes remained to him a
problem of individual taste, i.e. a purely subjective procedure. This
position drew critical comment from L. S. Berg. Berg also showed that
the Passarge concept of landscape excludes the need for the mutual
interdependence of its components.[22]

In evaluating the views of another representative of German landscape science, A. Penck, L. S. Berg wrote that Penck understood
landscape as a mechanical collection of various components, all
entirely unconnected and independent of one another; furthermore he
completely ignored the importance of soil.

West German landscape scientists even today often regard the
establishment of landscapes as a subjective process; many of them
also view landscape as a kind of entity embracing both nature and
man. In practical work concerning the physical-geographic differentiation of territories, however, they all proceed from their analyses of the
natural factors only.

The geographers of the German Democratic Republic adhere to the
concept of the objective nature of geocomplexes as units of a natural
classification of the earth's surface; they strive to re-examine the
theoretical principles of geography from the position of dialectical
materialism. Considerable attention has been devoted to detailed
physical-geographic regionalization of the East German territory.
Until recently this regionalization was the result of an in-the-office
comparison of various sources. Recently, however, the German

* German contributions to Western thought on regions have been rigorously
reviewed by Richard Hartshorne. (1939, pp. 426-541)

Academy of Agricultural Sciences has been conducting experimental field landscape mapping and to a large degree the mapping of land-scapes is based on the study of their morphological structure.[23]

In Poland interest in landscape science emerged largely as a result of the work of Soviet geographers. Since 1956–7 the Department of Physical Geography at the University of Warsaw has been conducting landscape studies under the direction of Professor J. Kondracki.* In 1959 Kondracki developed a classification of landscapes in Poland and compiled a landscape map at the scale of 1:1 000 000. According to Kondracki there has been a remarkable trend recently in Poland towards the study of natural complexes. This trend has attracted increasing attention among geographers.[24] The same trend charac-terizes contemporary Czechoslovakian geography.

Landscape science has remained undeveloped in most capitalist countries. One of the authors of the collection, *American Geography*, which summarizes the development of geography in the U.S.A. during the past century, admits that 'The concept of synthetic natural regions found little favor in America, where investigation focused on com-ponent elements of the natural environment.'[25] American geographers reveal considerable interest in regional geography, however, and the 'regional concept' is accepted as the cornerstone of geography; it is treated from a purely idealistic viewpoint.

As early as a quarter century ago an outstanding contributor to American geography, R. Hartshorne, wrote that regional geography integrates all the facts relating to a given territory and reflects both the natural and the socio-economic phenomena. Furthermore, 'the interest of the geographer is not in phenomena themselves, their origins or processes, but in the relations which they have to other geographic features, (i.e. features significant in areal differentiation).'[26] According to Hartshorne, geography studies a distinct territory or region (area), yet such areas have no real, objectively existing boundaries.

The same idea is encountered in the contemporary research of American geographers: 'a region is not an object, either self-determined or nature given. It is an intellectual concept, an entity for the purposes of thought, created by the selection of certain features that are relevant to an areal interest or problem and by the disregard of all features that are considered to be irrelevant.'[27] Moreover, American geographers are gradually intensifying their opposition to the concept of a 'region as objective reality ... and it is flatly rejected in this book'.[28] In addition, the author of this article thought it possible to speak about total regions by grouping together their physical, bio-

* (Kondracki, 1956)

logical and social components. These regions were regarded as the 'keystone of the geographic arch'.

This concept has little in common with the concepts of Soviet geography, which rests on the recognition of the physical-geographic differentiation of the earth's surface and on a fundamental distinction between natural and social factors. Very recently, however, American geographers began to reveal considerable interest in Soviet geographic theory. In particular, the monthly journal *Soviet Geography* published by the American Geographical Society in New York regularly publishes translations of papers by Soviet geographers. Many of these papers are devoted to the theory of landscape, problems of landscape research and the compilation of landscape maps.

An analysis of contemporary developmental trends in geography in the capitalist countries shows convincingly that they lead unavoidably to the recognition of geocomplexes and the acceptance of a science concerned with their investigation. This acceptance of landscape science, however, is very gradual and protracted; it is entirely unrelated to the theoretical background of bourgeois geography and arises out of practical needs: the needs of forestry, ecological evaluation of land, land planning, etc.* Thus the typology of forests in a number of countries, especially in Canada, employs a complex geographic approach and homogeneous areas of forest are regarded as complex natural units, corresponding to facies. In many countries including the U.S.A. and Great Britain, inventories of farming lands are being compiled and their soils qualitatively evaluated; maps of soil types are being published which distinguish sections which are more or less homogeneous with regard to the principal natural factors involved in agricultural production.† The principal criteria for the differentiation of types or classes of farming lands are the soils, relief, and sometimes the character of natural drainage and the climatic characteristics. Some of the most successful maps of land types approximate landscape maps in their content or, more precisely, landscape maps derived for a particular purpose. This activity, however, is not regarded as geographic, and even less so as landscape-scientific.

---

* A similar view is held by Davis in his review of the work of the Michigan Land Economic Survey. (1969, p. 41) There is, however, a long history of philosophical and methodological, if not theoretical, discussions in Western geographic literature. (Grigg, 1965, pp. 465-91)

† This is sometimes referred to as *land capability classification*. (Klingebiel and Montgomery, 1961)

2

General Principles of the
Physical-Geographic Differentiation of a Territory

Geographic zonality

The most complex geographic differentiation of the earth's surface and the diversity of its landscapes is due, in the final analysis, to the difference in the developmental history of its various parts. The direction and the rate of this development depend in turn on the relationship between the two major factors which determine the sources of energy for geographic processes: solar energy, and the internal energy of the earth.* Both these energy factors vary in time and space, and the character of their variations differs considerably. The specific manifestations of each of these factors in the geocomplexes on earth are responsible for the two most general principles of territorial differentiation: the principles of *zonality* and *azonality*.

Zonality, i.e. regular variation in all geographic components and landscapes with latitude from the equator to the poles, constitutes the best-known geographic law. We have shown in chapter 1 that zonality was first studied by the ancient Greeks, but that it did not achieve the status of a geographic law prior to the appearance of V. V. Doku-chayev's work during 1898–1900. Dokuchayev's students and his followers enriched the study of natural zones or landscape zones by providing new facts and generalizations concerning their nature.

The principal cause of zonality is the irregular distribution of solar energy at different latitudes. The angle of incidence of solar rays varies regularly with latitude and, as a result, the amount of solar energy received by a unit area of the earth's surface varies directly. In addition there are other factors which determine both the total amount

---

* Various terms designate these energy sources. In his elementary treatment of geomorphic processes, Thornbury discusses two sets: 'epigene' and 'exo-genous', both of which refer to processes which 'originate outside the earth's crust', while 'hypogene' and 'endogenous' both refer to processes 'which have their origin within the earth's crust'. (Thornbury, 1964, p. 34)

of solar energy reaching the earth and its distribution between the equator and the poles, thereby affecting the character of zonality. These factors include the distance between the earth and the sun, the size and mass of the earth, its diurnal rotation, and the inclination of its axis to the orbital plane.

S. V. Kalyesnik observed that there would be no zonality if the earth moved along an orbit as distant from the sun as Pluto. The earth would then receive 1600 times less energy than it receives now and its entire surface would be a uniform ice desert. Zonality also depends indirectly on the earth's size; a reduction in size and mass of our planet would eventually result in the loss of the atmosphere, which would in turn eliminate zonal phenomena. The inclination of the earth's axis to the orbital plane is responsible for the uneven supplies of solar energy from one season to the next which complicates zonality by intensifying the contrasts between zones and increasing their number. Finally, the diurnal rotation of the earth adds still further to the complexity of zonal phenomena. A major role is played by the Coriolis force which is responsible for the deviation of moving bodies including atmospheric masses, leading to a substantial redistribution of heat and moisture.

Solar energy is virtually the sole source of the physical, chemical and biological processes on the surface of our planet. The intensity of these processes is determined directly by the amount of heat and light received from the sun, and for this reason the processes must be zonal in character. Thus the phenomenon of *zonality* on the earth is entirely due to solar system factors. The forms by which it manifests itself, however, are determined by the nature of the geographic envelope, which actively transforms all these external influences. It would be more accurate, therefore, to say that solar system factors control only the gross influences responsible for zonality. In the specific conditions of the geographic envelope, with its complex structure and diversified material composition, zonality becomes a concrete phenomenon. If the earth's zonality were a simple reflection of astronomical factors, it would have a mathematically regular character. The changes in heat conditions and in all other natural phenomena would lead to the development of regular circular zonal bands; these would gradually merge into one another, yet they would have parallel boundaries. In other words, we would be dealing not with geographic but with *astronomical zonality*.

In fact, landscape zones form a very complex pattern on the earth's surface. They are often broken and hardly ever parallel in orientation, their boundaries are irregular and the transition from one zone to another is sometimes sudden and sometimes extremely gradual. (The reasons for these phenomena will be discussed in detail below.) Furthermore, zonality manifests itself differently in every geographic component.

The uneven heating of the earth's surface by solar rays at different latitudes is reflected most directly in atmospheric phenomena: the zonality of barometric pressure, atmospheric circulation, and the hydro-thermal properties of atmospheric masses. The principal zonal types of air masses—arctic, boreal (temperate latitudes), tropical and equatorial—determine, as it were, the primary schema of broad climatic belts. The seasonal variations in atmospheric circulation between the four principal zones produce three intermediate zones in which two different types of air masses interchange seasonally. This results, according to B. P. Alisov, in seven major climatic zones for the northern hemisphere: the arctic and subarctic, temperate, subtropical, tropical, subequatorial and equatorial zones. Moreover, regular variation in all the major climatic indices is observed between the equator and the poles. (Fig. 2)

Another important consequence of the uneven distribution of the solar heat supply on the earth and of the existence of a system of circulation zones is the zonality of moisture conditions. According to O. A. Drozdov,[1] the moisture content of air masses in general, disregarding certain details, increases steadily from the poles towards the equator. Thus an air column 7 km high over the Arctic contains no more than 5 mm of moisture in January and less than 10 mm in July; over a forest belt in the temperate zone in the northern hemisphere the corresponding amounts are approximately 5 mm and 25 mm respectively, over subtropical deserts approximately 10 mm and 25 mm, and over the equatorial zone atmospheric moisture exceeds 40 mm the whole year round.

Atmospheric moisture, however, does not by itself determine whether a climate is wet or dry. Even the amount of rainfall is not a determinant, since the effect of the same amount of rainfall may be completely different depending on the conditions of its utilization. For example, at 200 mm annual precipitation over the tundra, soils become water-logged and considerable runoff takes place, whereas in semi-desert areas the same amount of rainfall would be entirely inadequate for the soil and the vegetation, and the runoff would be virtually nonexistent.

Conclusions about moisture conditions must be based on a comparison between the moisture received from the atmosphere (i.e. rainfall) and the potential moisture loss for given climatic conditions, (i.e. evaporativity).[2] The first determination of the relationship between annual rainfall and evaporativity for the geographic zones of European

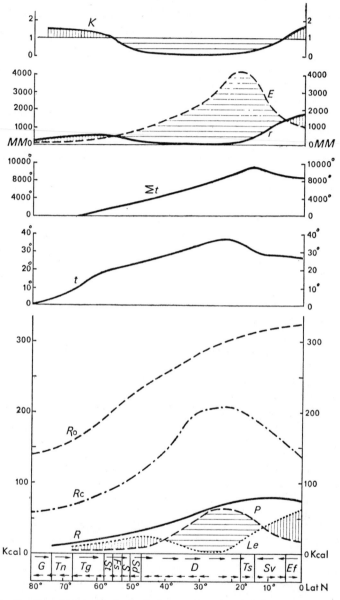

Fig. 2 Zonal variations in the basic heat and moisture indices on land in the northern hemisphere

Landscape zones: *G*, glacial; *Tn*, tundra; *Tg*, tayga; *St*, subtayga; *Fs*, forest

Russia was made by G. N. Vysotski in 1905. Much later N. N. Ivanov calculated the value of this index, which he called the moisture coefficient, for many points round the globe. If the coefficient equals 1, representing the situation where total annual rainfall and evaporativity are equal, it indicates optimal conditions of atmospheric moisture for most land uses; a lower value indicates insufficient moisture and a higher value excessive moisture.

Changes in the moisture coefficient and in absolute rainfall with latitude are not all in one direction and do not always conform to changes in thermal conditions. (Fig. 2) The zonality of moisture conditions has, as it were, a rhythmic character, with two peaks—in the equatorial and temperate-polar latitudes—and a minimum in subtropical latitudes. The causes of this phenomenon must be sought in the general circulation of the atmosphere, which is the major determinant of rainfall distribution. Atmospheric circulation is not only a function of solar radiation but also of the earth's rotation. If the earth were to cease rotating, air streams produced by the differences in the heating of the polar and equatorial zones would follow the meridional direction from the poles (areas of maximum pressure) towards the equator (the area of minimum pressure). In the upper layers of the atmosphere the direction of the air streams would be opposite, owing to the spread of the rising air masses over the equator.

The diurnal motion of the earth, however, introduces significant variations into this simple system. In particular, high pressure zones

---

steppe; *S*, steppe; *Sd*, semidesert; *D*, desert; *Ts*, tropical semidesert; *Sv*, savanna; *Ef*, equatorial forest.

The graph shows a cross-section along a typical meridian, i.e. in a temperate continental climate with lowland relief and a regular latitudinal distribution of zonal boundaries (approximately along 50° E. in eastern Europe and central Asia and along 20° to 30° E. in northern Africa). *Ro*, solar radiation in the absence of atmosphere; *Rc*, real total radiation on the earth's surface; $R = Le + P$, radiation balance (difference between the amount of radiation absorbed by the earth's surface and the amount radiated by it); *Le*, radiation heat loss due to evaporation from the earth's surface; *P*, turbulent heat flux from the underlying surface into the atmosphere. All these indices are expressed in $kcal/cm^2$ and are shown in a single scale. Vertical hatching indicates the excess heat loss by evaporation over the turbulent exchange between the earth's surface and the atmosphere; horizontal hatching designates the excess of turbulent exchange over evaporation losses.

t, mean air temperature in July; Σt, mean annual sum of temperatures over a period, with average daily temperatures in excess of 10° C; *r*, mean annual rainfall in mm; *E*, mean annual evaporation in mm; $K = r/E$, moisture coefficient. Vertical hatching indicates excessive moisture and horizontal hatching indicates insufficient mosture. Long arrows show the direction in which heat supplies in landscape zones increase and short arrows show the direction in which moisture supplies increase.

due not to thermal but to dynamic factors are created in subtropical latitudes, approximately along the 30th parallel; the air flowing away from the equatorial zone as anti-trades gradually turns in an easterly direction and precipitates its moisture, causing a pressure increase at ground level. The prevalent downward movement of the air in this zone hinders rainfall. In the equatorial zone, on the other hand, the prevalence of convection (rising) streams of very moist air leads to intensive precipitation. The substantial rainfall in temperate latitudes is associated with westerly movement, again dynamic in origin, and cyclonic activity. In polar latitudes, the amount of rainfall is reduced owing to increased pressure and a reduction in atmospheric moisture; but since evaporativity reduces in a polar direction at a faster rate than rainfall, the zone receives excessive water. In its variation with latitude, evaporativity is closely associated with heat resources, since the shape of its curve differs from the shape of the rainfall curve and closely resembles the curve of radiation indices. (Fig. 2)

The amount of active moisture is the next most significant factor which determines the character and the intensity of geographic processes, e.g. runoff, weathering, soil formation, exogenous transformation of relief, and the biological activity of biocenoses. Accordingly, in discussing climatic zonality factors it is necessary to distinguish the role of solar energy from that of atmospheric moisture. Both these factors always occur together as an indirect cause of zonal phenomena in other components of the landscape.

There is a close interdependence of heat supply and landscape moisture. This interdependence is expressed not only in the above-indicated dependence of moisture on the distribution of solar radiation around the globe, but also in the opposite relationship, i.e. in the definite effect of moisture conditions on heat conditions at ground level. Thus in zones with insufficient moisture, heat losses in evaporation are negligible and the overwhelming proportion of radiation heat via turbulent heat exchange is used in heating air. By contrast, both in wet equatorial zones and in temperate and polar zones with excessive moisture, the overwhelming proportion of heat supply is used in evaporating atmospheric moisture and, as a result, a characteristic thermal anomaly is found to exist in dry subtropical and tropical latitudes: in summer the highest air temperatures are found in these zones. Moreover, the amount of solar energy reaching the ground depends on the humidity of the climate: in zones with excessive moisture a high proportion of solar rays are reflected and diffused by the clouds, so that a sharp peak of total solar radiation is actually observed not on the equator but between 20° and 30° latitude, i.e. in the zone of dynamic barometric maximum, having a dry climate and predominantly clear skies. For these reasons, the real distribution of solar radiation at ground level differs considerably from the theoretical distribution which would obtain were there no intervening atmosphere. (Fig. 2)

It follows that the cumulative variations in heat supply and moisture do not reveal a simple pattern which we would expect on the basis of the gradual increase in the angle of incidence of solar rays from the poles towards the equator. Latitudinal gradients of changes in heat resources and atmospheric pressure vary both directionally and in amount.

Climatic zonality affects hydrological processes and water balance sharply. Table 1, compiled by M. I. Lvovich,[3] lists the principal water-balance indices for certain drainage basins lying in different geographic zones. (The term *total moisture content* in the soil refers to that portion of atmospheric rainfall which is not taken away by the surface flow and is eventually lost in evaporation and in supplying groundwater.)

The zonality of groundwater was discovered in 1914 by a student of V. V. Dokuchayev, P. V. Ototski. Later V. S. Ilin, B. L. Lichkov, I. V. Garmonov, G. N. Kamyenski and many others studied this problem. It can be regarded as firmly established today that climatic zonal factors affect the conditions in which groundwater is formed, its heat balance, mineralization and ionic content. The effect of zonality on the depth of groundwater is less obvious, since it is disguised by relief, the diverse lithology of rocks, and other local factors. If we were to compare, however, sections of different zones with raised, slightly sloping watershed localities, we would find that from the tundra to the desert the watertable falls at a regular rate with the increased dryness of climate.

It is necessary to consider separately the zonality of geochemical processes, which plays an important role in the evolution of the geographic envelope and its landscapes.[4] The major source of energy for geochemical processes is solar radiation which organisms convert into chemical energy. The medium in which most chemical reactions take place is water. It is obvious that geochemical processes, which include intensive dissolution of rock, oxidation and reduction reactions, migration of aqueous solutions in the soil and bedrock, and the removal of chemical elements from the weathered crust, must be strongly affected by latitudinal zonality.

These processes, in turn, influence the formation of secondary minerals and the development of different types of weathering mantles.

Table 1

Geographical zone	River basin	Rainfall (mm)	Runoff (mm)			Evaporation (mm)	Total soil moisture (mm)	Runoff coefficient
			Total	Including				
				Surface	Under-ground			
Tundra	Amguyema	350	255	247	8	95	103	0·73
Tayga	Vym	530	314	238	76	216	292	0·56
Tayga in permafrost	Olyenyok	400	195	185	10	205	215	0·49
Mixed forest	Klazma	580	210	160	50	370	420	0·36
Forest steppe	Myedvyeditsa	410	78	62	16	332	348	0·19
Steppe	Sal	350	30	26	4	320	324	0·09
Semidesert	Turgay	200	10	9	1	190	191	0·05
Semidesert	Zak	195	11	10	1	184	185	0·06
Savanna	Shari	990	63	53	10	927	937	0·06
Subtropical forest, mediterranean type	Dorn	550	68	62	6	482	488	0·18
Tropical monsoon forest	Murakshi	1500	550	440	110	950	1060	0·37
Tropical rain forest	Nyan Mou Tsyan	2000	900	650	250	1100	1350	0·45

Geochemical processes determine the alkaline-acidic properties of water solutions in the landscape, the degree of mineralization and the ionic content of soil and underground, river and lacustrine waters, the content of base exchange ions in the soil, the subtraction-accumulation ratio of humic substances, etc.

An important geochemical index to the various natural zones are the so-called typomorphic chemical elements,* i.e. elements which in given conditions possess maximum mobility and chemical activeness and which determine the chemical properties of aqueous solutions. In the tundra geochemical processes are weakened, the core of weathering is mainly fragmental, and the primary minerals change very little. The typomorphic elements H^+ and Fe^{++} determine the high acidity of soil solutions. The geochemical processes in the tayga involve the active role of organic (humic) acids. The downward transport of solutions predominates; calcium, magnesium, sodium and other elements are removed; a siallite mantle of weathering develops (because of Si and Al) which consists mainly of mixtures of hydrates of SiO_2, Al_2O_3 and Fe_2O_3 and their derivatives alumo- and ferrosilicate. The principal typomorphic ion in the tayga is H^+. It determines acidity conditions and the lack of calcium in soils.

In the steppe the alternation of downward and upward flow in the soil-bedrock layer is observed; here humic acids play an important role in the breakdown of primary rock. The products of weathering have a loessic character; they are predominantly siallitic in content but they are also enriched with calcium and magnesium carbonates.† Calcium is the principal typomorphic element. The presence of calcium in the absorbing complex determines a neutral or weakly alkaline reaction of soil solutions, assists the flocculation of colloids, and increases the fertility of soils.

The desert zone is characterized by the upward flow of water solutions in the soil-bedrock layer and the role of organisms in geochemical processes is very slight. In this zone argilloarenaceous products result from weathering and the chloride and sulphate salts of sodium, calcium and magnesium are accumulated. The typomorphic ions are Na^+, Cl^- and also, in part, Ca^{++}. Soil solutions have an alkaline reaction.

---

* Typomorphic chemical elements are discussed with regard to landscape classification by Perelman (1961). A table is provided in which landscape types are compared to typomorphic elements of water migration.

† In this book reference is occasionally made to carbonate, carbonaceous, or non-carbonate conditions, especially with reference to soils. The term 'carbonate', etc. in Russian usage does not always refer to material containing $CaCO_3$. For this reason it is not translated as 'calcareous'. See *Soil-Geographical Zoning of the USSR* (1963).

50   *Landscape Science*

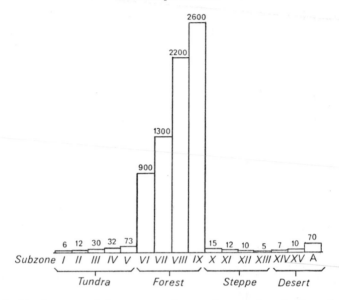

Fig. 3   Total amount of the overground vegetative mass in upland zonal plant communities in European U.S.S.R. and in Soviet central-Asia, shown in centner (100 kg)/hectare (10m²) (after E. M. Lavrenko, V. N. Andreyev and V. L. Leontyev)

Subzones: *I*, arctic semidesert; *II*, arctic tundra; *III*, northern tundra; *IV*, southern tundra; *V*, thin-forest tundra; *VI*, northern tayga; *VII*, middle tayga; *VIII*, southern tayga; *IX*, broad-leaved forest; *X*, grassland steppe; *XI*, typical mottley-grass, soddy steppe; *XII*, typical soddy-grassland steppe; *XIII*, semi-scrub desert and soddy-grassland steppe; *XIV*, steppe-like (northern) and typical semi-scrub desert; *XV*, ephemeral semi-scrub (southern) desert; *A*, black haloxylon.

The wet subtropical and especially the equatorial zones are characterized by a biogenic cycle of maximum intensity and by the water migration of chemical elements. We find here an intensive subtraction of calcium, magnesium, sodium and potassium, as well as silicon; yellow-soil, red-soil and lateritic mantles of weathering, enriched with the hydrates of Al_2O_3 and Fe_2O_3 develop. The typomorphic ion is H^+. Soil solutions have a weakly acidic to weakly alkaline reaction.

Geographic zonality is most sharply reflected in the organic world. It is not due to an accident that landscape zones were in most cases given their names on the basis of their characteristic types of vegetation. Zonality is reflected in the distribution of the principal forms of plants and animals, in the structure of the biocenoses, their species composition and diversity, and in many other typical characteristics. It

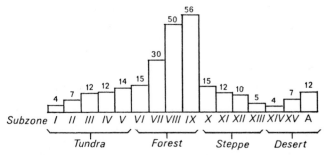

Fig. 4 Annual increase in the overground vegetative mass in upland zonal plant communities of the U.S.S.R., shown in centner (100 kg)/hectare (10m²) (after E. M. Lavrenko, V. N. Andreyev and V. L. Leontyev)

Subzones: *I*, arctic semidesert; *II*, arctic tundra; *III*, northern tundra; *IV*, southern tundra; *V*, thin-forest tundra; *VI*, northern tayga; *VII*, middle tayga; *VIII*, southern tayga; *IX*, broad-leaved forest; *X*, grassland steppe; *XI*, typical mottley-grass, soddy steppe; *XII*, typical soddy-grassland steppe; *XIII*, semi-scrub desert and soddy-grassland steppe; *XIV*, steppe-like (northern) and typical semi-scrub desert; *XV*, ephemeral semi-scrub (southern) desert; *A*, black haloxylon.

should be emphasized that the effects of zonality spread even to micro-organisms, and in particular to the bacteria content of soils.

Rather than characterize the various zonal types of plant cover, data concerning the supply of plant mass and its productivity in various zones of the U.S.S.R. is presented. (Figs 3 and 4) These quantitative indices provide more than just an illustration of geographic laws; a detailed study of the productivity of the plant cover in differing geographic conditions, taking account of its seasonal dynamics, structure and chemistry, is of very great practical and theoretical significance. This is also essential for the study of the biogenic cycle of substances and the role of vegetation in soil formation and land-scape development. Unfortunately adequate information in this area is still lacking.

The zonality of the soil mantle is functionally bound, as already shown by V. V. Dokuchayev, with the zonal variation in geographic soil formation factors: climate, water flow, the organic world, and geochemical processes. The zonal distribution of the principal genetic soil types is so well known that there is no need to discuss it in detail.

The relief of the earth's surface is often called the azonal component as if to emphasize that it is not subject to the law of geographic zonality. This idea, however, must be viewed as erroneous. Geo-graphic components cannot, in general, be divided into zonal and azonal, since, as will be seen later, any such component combines both zonal and azonal characteristics. So far we have only touched on the former. It is well known that relief develops as a result of the combined action of endogenous factors, namely the tectonic, and exogenous factors associated with processes taking place on the surface of the earth under the direct or indirect action of solar energy (weathering, erosion and accumulation, winds, glaciers, etc.). All exogenous processes are zonal in character and for this reason the shapes they impart to relief are classified as 'sculptural' (as against 'structural', produced by tectonic and volcanic processes) and must be distributed over the earth's surface in zones. Ya. S. Edelshteyn, M. B. Gornung and D. A. Timofyeyev, and M. V. Karandyeyeva, as well as other Soviet and foreign geographers have devoted considerable attention to these processes.

In glacial zones (Arctic and Antarctic) the principal forms of relief are determined by current glaciation. Glaciers are directly involved in moulding surface relief (e.g. highland glacier plateaus, ice-caps, glacial streams and barriers). Frost weathering and periglacial accumulation prevails in areas free of ice cover. In the tundra specific surface forms develop as a result of frost weathering and permafrost; these include sinkholes, swelling knolls, patterned ground, etc. Erosion in this zone is retarded, lateral erosion predominating, and as a result the river channels are shallow but wide.

In the tayga the intensity of erosion is slight, since runoff is strongly retarded by the forest cover; rounded forms with gentle slopes predominate. Processes leading to the overgrowing of lakes are strongly developed and assist the relief-levelling tendency. Specific meso- and micro-forms of relief characterize swamps, which occupy large areas in this zone. The tayga is also characterized by typical relict forms of glacial and fluvio-glacial accumulations (e.g. moraines and outwash plains).

The forest-steppe zone combines the geomorphological properties of each constituent zone, forest and steppe. The latter zone is distinguished by (1) intensive erosion, mainly gully-type, assisted by the zonal characteristics of the steppe (and the forest as well); (2) rapid thawing of snows in the spring and frequent torrential rains, both of which produce intensive surface runoff; (3) easily eroded loess and loess-type soils; and (4) the absence of forest vegetation. In lower-lying and poorly-drained areas of the forest-steppe and the steppe zones, minor depressions and sinkholes are widely distributed.

In semidesert and, in particular, desert zones with arid climates, sparse plant cover and negligible runoff, erosion processes are sharply reduced and only the effects of seasonal streamflow are of any significance. In areas with loose, sandy soils the predominating relief-forming factor is wind activity. These processes explain the typical features of desert relief: the diversities of aeolian forms, takyrs,† drainless solonchak depressions, proluvial* debris cones (in foothill areas) and the accumulation of rock-waste products by mechanical erosion.

Rather than treat in detail the geomorphological features of other zones (e.g. savanna, tropical monsoon forest, etc.) only the major zonal features in relief formation characteristic of equatorial forest will be noted. Here intensive chemical erosion results in the formation of a siallitic weathering crust which may reach a thickness of some tens of metres. Abundant water supplies produce slumping and mud-avalanche flow of loose soils, augmented by surface runoff, bog formation and intensive accumulation of alluvium. Erosion may attain vast dimensions when accompanied by the destruction of forest.

The law of zonality also leaves its mark on such specific phenomena as karst, or the formation of sea coasts, which occur in different zones. Thus karst formation proceeds most intensely in mediterranean zones and in wet tropical zones. In temperate forest zones we do not encounter a similarly complete development of karst forms. In the arctic and tundra zones the conditions for karst formation are even less favourable; in desert zones karsts are rare and the karst-formation processes are combined with intensive mechanical weathering.

The deposition of sedimentary material which occurs in the geographic envelope and is subject to its laws has a zonal character, as it depends directly on thermal conditions, moisture, runoff and the organic world. Numerous facts confirm this assertion and examples of the zonal distribution of various sedimentary rocks can be readily cited. As early as 1932, V. V. Alabyshev described the zonality of lacustrine deposits. N. M. Strakhov,[5] a leading contributor to the theory of sedimentary rock formation, established three principal zonal types of lithogenesis: glacial, arid and humid. The first type of lithogenesis develops in conditions of continental glaciation of the polar zones and is characterized by the mechanical destruction of rocks and the transport and deposition of unsorted material. The arid type characterizes deserts, with their specific climatic and geochemical

---

* 'Takyrs' are soils found on flat plains; when dry they are hard with polygonal cracks, when wet they are gluey, making cross-country mobility extremely difficult.

† 'Proluvial debris' refers to material found at the base of slopes and transported there by downwash and overland flow. Isachenko also uses the term 'colluvial' to mean material transported by gravity and/or frost action. The terms seem to be used synonymously in the Moscow edition, however, and it is difficult to determine if a distinction is intended. A more general term found in the book is 'deluvial'.

conditions; here clastic rocks and salts constitute the major material for accumulation. Humid lithogenesis plays the most important role in the formation of the stratisphere and is characterized by the great diversity of sedimentary processes. These in turn affect differently the wet areas of the cold, temperate, subtropical and equatorial zones. Strakhov emphasized that such a geographic differentiation of sedimentary formation typifies not only the present geological epoch but has existed over the entire course of geological history.

The inevitable consequence of the interdependent zonal variations found in various geographic processes and in individual geographic components is the zonality of the earth's landscapes. Landscapes in regular patterns constitute a system of landscape zones, each of which forms an independent geographic geocomplex of a higher order. The principles governing the differentiation of landscape zones, the determination of their boundaries, and their comprehensive description, constitute the tasks of physical-geographic (landscape) regionalization. In this chapter we are only concerned with showing by various examples the general character and the diverse manifestations of zonality in the geographic envelope.

It is necessary to realize, on the other hand, that the effects of geographic zonality do not go beyond certain limits. These effects are clearly exaggerated in the relatively recent attempts of certain geographers to invoke climatic zonality in explaining the distribution of mountain ranges and orography of continents, and in accounting for the regularities of tectonic movements.

Geographic zonality manifests itself directly in those areas where solar radiation interacts with the material of the geographic envelope, i.e. within the relatively narrow 'film' on the surface of land and ocean. The effect of zonality gradually attenuates towards the outer boundaries of the geographic envelope, yet the indirect consequences of the processes associated with zonality which occur at the interface between the lithosphere, hydrosphere and the atmosphere may spread a long way in both directions.

These indirect effects of zonality manifest themselves in an especially complex way in the lithosphere. They are manifest at substantial depth owing to the penetration of air and water along rock fissures, and to the deposition of sedimentary rocks in which solar energy is accumulated. Investigations by N. I. Tolstikhin have shown that, in contrast to prevailing opinion, zonality manifests itself in the properties of deep artesian waters.[6] It is expressed primarily in the temperature changes of artesian waters with latitude: the depth of isothermal horizons (i.e. horizons with identical temperature) increases from south to north. For example, artesian waters with 20° C temperature are found in the Kara Kumy and Kyzyl Kumy at a depth of only a few tens of metres, while in south-western Siberia they lie at a depth of 650 to 700 m, and in central Yakutia at 900 m plus. Artesian waters with 0° C temperature are found in the Transbaykal at a depth of less than 100 m, in central Yakutia at 150 to 200 m and only at a depth of 400 to 600 m in the Khatanga Basin.

Corresponding changes in the principal properties of artesian waters are also found; along identical depths there are variations in mineralization, in the amount of dissolved salts, and in biological conditions. Fresh artesian waters are found only in zones of excessive and adequate moisture and in such areas they can reach a depth of 200 to 300 and even 500 m, but in the far north these waters are frozen. In dry-climate zones the horizons of fresh artesian waters are either very thin or entirely absent. The above indicates that water supplies along the very deep levels of the lithosphere (1000 m and lower) are associated, through descent and evaporation, with processes occurring on the surface. Geographic zonality is manifested on the surface of the hydrosphere by variations in temperature, salinity, gas content of the water, and the dynamics of the surface horizons in the oceans, as well as in the organic world (including the plankton) by its mass, specific composition, etc. The effects of zonality reduce with depth but can still be distinguished on the ocean bed where they determine the character of the bottom mud, the origin of which is essentially organic. More clearly, zonality is reflected in the landscapes of the continental shelf, especially in areas exposed to the direct effect of processes occurring on the adjacent sections of continents.

In the troposphere zonal differences decrease with height; this lessening is assisted by the homogeneous composition and the great mobility of the air. A type of zonality also characterizes the upper layers of the atmosphere. At levels of the order of 400 to 500 km, oxygen molecules split off from nitrogen molecules and the atmosphere contains substantial quantities of ionized atoms of these gases. Data obtained by the Soviet artificial satellites indicate that at these altitudes the content of atomic oxygen and nitrogen ions is higher in high latitudes than in low latitudes. This zonality has nothing to do, of course, with that extending over the lower layers of the atmosphere, and cannot be considered a geographic zonality. The zonality of the troposphere results from its interaction with the surface of land and ocean; the ground levels of the atmosphere receive heat from below, hence air temperature reduces with altitude. The properties of the upper layers of the atmosphere extending beyond the geographic envelope are formed by the direct interaction between the solar radiation flux and atmospheric gases. This interaction is responsible, in particular, for the rise of temperature with altitude at these levels. A distinguishing characteristic of the upper layers of the atmosphere is also short-wave radiation, which is capable of destroying life. It follows that zonality, as a specific natural phenomenon inherent in the geographic envelope, manifests itself directly or indirectly only in that envelope.

Finally, it is necessary to note the effects of zonality on the economic activity of man. Although in the past many geographers, including Dokuchayev, grossly exaggerated the effect of this factor, it is equally wrong to reject it entirely. Zonality affects agriculture particularly strongly, but it also affects building construction and public health. Dokuchayev's great achievement, once we disregard his mistaken attempt to extend the effects of the zonal law to the existence, cultural achievements and other aspects of human society, is his pioneering attempt to give a scientific foundation to the essential task of the independent application of agro-technological and land-reclamation measures in the natural zones of Russia.

Contemporary and historical factors associated with zonality, and the genetic principle in the study of geographic zones

It is often said that geographic zonality is a function of climate, or that climate is a factor in zonality. Geographers and climatologists have devoted considerable effort to the task of deriving a synthetic quantitative index of climatic factors and to defining the relationship between these factors and the zonal distribution of various types of vegetation, soils and landscapes. During the last five to six decades many such indices, e.g. hydrothermal coefficients generally representing various empirical combinations of mean air temperatures and amounts of rainfall, have been proposed. The majority of these lack any substance and are of no scientific interest.[7] A number, however, are of some value and find application in the analysis of zonal laws in geography. They include the Vysotski-Ivanov moisture coefficient, (K) already discussed earlier, which expresses the ratio of total annual rainfall (r) to annual evaporativity (E).

A comparison of the isolines of the mean annual moisture coefficient over the territory of the Russian plain with the boundaries of landscape zones (Fig. 5) reveals the following ratios: forest zones, as well as the tundra, show coefficients in excess of 1, forest-steppe from $1 \cdot 0$ to $0 \cdot 6$; steppe from $0 \cdot 6$ to $0 \cdot 3$; semidesert from $0 \cdot 3$ to $0 \cdot 12$; deserts from $0 \cdot 12$ to $0 \cdot 8$.[8]

M. I. Budyko and A. A. Grigoryev hold the view that the value of

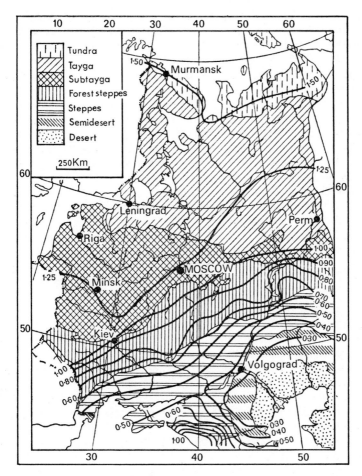

Fig. 5   Isolines of the mean annual moisture coefficient over the territory of the Russian plain

the total annual radiation balance (R) and annual rainfall (r) determine all processes characterizing the external levels of the geographic envelope (i.e. those parts of the envelope which are subject to the direct effect of solar radiation). The ratio of the total annual radiation balance to the total amount of heat necessary for the evaporation of the total annual rainfall (i.e. R/Lr where L equals potential heat of evaporation, approximately equal to 0·6 Kcal/g) is designated by Budyko and Grigoryev as the radiation index of aridity; they assume that the boundaries of landscape zones must coincide with certain specific values of this index. And in fact, in very broad outlines, such correlation exists. In the tundra the radiation index of aridity is less than ⅓; in the forest zones of the temperate, subtropical and equatorial regions ⅓ to 1; in the steppe 1 to 2; in semideserts 2 to 3; and in the deserts 3 and more.

According to Budyko and Grigoryev, the index range from 0·8 to 1·0 indicates the optimal ratio of heat to moisture; lower values indicate increasing excess moisture and higher values increasing aridity. Since the optimum value of this index recurs in several zones (e.g. broad-leaved forests and forest-steppe in the temperate zone, wet subtropical forests and equatorial forests) the authors derive a 'periodic law of geographic zonality'.[9]

It is not difficult to show that the radiation index of aridity resembles, in its physical sense, the moisture coefficient of Vysotski-Ivanov. If we divide both the numerator and the denominator of the fraction R/Lr by L, we obtain precisely the ratio of the maximum possible, in the given radiation conditions, evaporation (i.e. evapora-tivity) to total annual rainfall, i.e. an inverted Vysotski-Ivanov coefficient. The difference between the two is very small; for Budykov and Grigoryev the criterion of evaporativity is the radiation balance, for Vysotski and Ivanov that criterion is provided by temperature and atmospheric moisture content. This leads to some differences in the estimated degree of atmospheric moisture.

Relatively close correlation is found between the boundaries of landscape zones and the isolines of the e/E coefficient, where e equals actual total annual evaporation and E equals evaporativity. (Fig. 6) This coefficient may be used as the index of the intensity of moisture exchange between the earth's surface and the atmosphere. If we take into account the fact that the overwhelming proportion of total evaporation results from transpiration, then the ratio e/E shows approximately the extent to which solar energy is effectively utilized by the plant cover in the creation of the biomass (i.e. the accumulation of dry mass in plants is proportional to the amount of transpiration). Over the Russian plain, heat resources are most efficiently utilized for transpiration and assimilation in the southern subtayga (mixed-forest zone) and in the northern forest-steppe (in the subzone of broad-leaved forests) where actual evaporation is 0·8 to 0·9 of the theoretical value. To the north and to the south of this belt the value e/E decreases uniformly—to the north because of insufficient heat, and to the south because of insufficient moisture—and the intensity of moisture exchange and the degree of utilization of solar heat by the plant cover falls correspondingly.

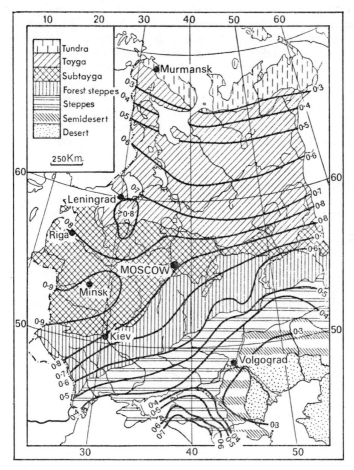

Fig. 6 Relationship between the mean total annual evaporation and evaporativity on the Russian plain

These conclusions are in agreement with the data on the accumulation and the productivity of the plant mass, shown in figures 3 and 4.

In evaluating the various indices of heat and moisture supply for landscapes, it must be kept in mind that they reveal only an approximate correlation with the boundaries of natural zones. In fact it is barely conceivable that one could select a universal quantitative index of climate which would more or less accurately coincide with zonal boundaries. To begin with, no single coefficient could express the entire complex set of climatic phenomena. Secondly, because climate is not the only factor in zonality, it would be a mistake to attribute to it the determining role in the formation of landscape zones.

The interdependence between climate and other geographic components is relatively complex. In some situations temperature conditions may have a determining effect; thus the distributions, for example, of many plant communities depend primarily on the ability of the plants to survive in winter conditions, which in turn is determined not so much by the mean as by the extreme (minimal) temperatures. In other cases the distributions of plant communities, as also of soils, are associated with perennial variations in moisture conditions or with the duration of periods when the atmospheric and soil temperature remains above a certain level (e.g. above $10°$ to $15°$ C), or with the altitude and the duration of the winter snow cover, etc. In any case, a particular index of climate may only be considered significant when related to all the other climatic phenomena. Thus the total annual radiation balance (R) can hardly be considered a unique or universal index of the energy resources of a landscape, if only because evaporation may also take place when the radiation balance is negative owing to heat inflow by advection.

A defect common to all these indices (r/E, R/Lr, e/E) is their remote character and their inability to express the absolute intensity of processes and phenomena. Each of these indices has an identical value in tayga and in equatorial forest zones, even though the rainfall intensity or moisture exchange is very much greater in the latter case than in the former.

A particularly important defect of the various universal climatic coefficients is their inability to reflect seasonal changes in heat and moisture conditions. Accordingly it is necessary to be extremely careful in using them in different kinds of geographic comparisons. Let us consider one example. The analysis of spatial variations in the moisture coefficient for the radiation index of aridity may lead us to conclude that the entire rainfall in the steppe must evaporate, and that there is no runoff whatever. In fact the situation is quite different, since heat conditions and rainfall vary considerably with the seasons so that in certain periods of the year (e.g. during the thaw or torrential rains) the steppe is characterized by excessive moisture which cannot be evaporated and is channelled off along watercourses.

Climatic coefficients, similar to those discussed above, are without doubt very important for the description of the heat and moisture conditions of various geographic zones and for the evaluation of the role of climate in landscape. However, a comparison of zonal boundaries with the values of this or that individual coefficient should not be treated as a substitute for the analysis of landscape and climate relationships, which would account as fully as possible for every significant factor on a seasonal basis. One can hardly draw reliable conclusions from the study of correlations between one or two indices and zonal boundaries.

As noted earlier, climate, despite its obvious importance, is certainly not the only factor determining zonality. An investigation of climatic factors cannot adequately explain all the principal features of the current physical-geographic differentiation of zones, their structure and the location of their boundaries.

The concept 'climatic factors of zonality' has a fairly arbitrary character. Strictly speaking there exists only one primary and autonomous climatic factor of zonality: the solar radiation along the upper boundaries of the atmosphere. Also ideally zonal is the so-called solar climate, i.e. a fictitious climate wholly independent of geographic factors and determined exclusively by the latitudinal distribution of solar radiation. But when solar rays enter the atmosphere and reach the earth's surface they are subject to complex transformations. The radiation balance does not constitute an independent factor but, as it were, a secondary element—as a result of the transformation of the solar radiation flux in the geographic envelope. Its value depends on many factors including the condition of the atmosphere, in particular cloud cover, and of the earth's surface (e.g. the distribution of land and water, relief, vegetation, snow mantle, etc.).

Moisture is a product of a broad spectrum of factors; it is a function of both solar radiation and of numerous geographic components. Accordingly, the climate itself is to a considerable degree a product of interaction of geographic components, and the distribution of climates is subject not only to zonal but also azonal factors. Thus zonality is not merely a simple manifestation of contemporary climate. As early as 1923, B. A. Keller wrote that the formation of zones was accompanied by simultaneous changes in climate, soils and vegetation, each strongly affecting the other; for this reason the isolation of climate as a major determinant is purely arbitrary.

Each landscape zone emerged slowly and developed over many thousands of years; each therefore has its own individual history. During its period of development the climate did not remain constant. The organic world and relief also underwent changes. Climate, as is well known, is the most dynamic component, and is subject to continuous change including cyclic change. Variations in other components, vegetation, soil and particularly relief, lag climatic changes. L. S. Berg concluded from this that landscape zones need not fully coincide with climatic zones.

The correspondence of all geographic components with contemporary climate has a relative character. Even in vegetation, which is highly responsive to all variations in the geographic environment, some contemporary zonal features cannot be attributed to climate. It is necessary to account for the origin and the age of the zone and its plant communities. Local sparseness of flora and fauna is often due not so much to a somewhat unfavourable zonal climate as to the immaturity of the landscape. For example, the organic cover of the Eurasian tayga or the eastern European zone of mixed forest (sub-tayga), is much poorer than the analogous, yet considerably older, far-eastern subtayga.

Landscape zonality therefore constitutes a mechanism or a genetic category and zonality is also the oldest geographic mechanism. Individual zones obviously existed in the Archean era, i.e. at least two billion years ago. Even then the thermal conditions on the earth were determined by the influx of solar heat and its latitudinal distribution; at that time sedimentary rocks were already forming and the atmosphere and the hydrosphere were in evidence. Since then, however, zonality has undergone complex development so that contemporary zones have very little in common with those of the Archeozoic and the Paleozoic eras.

The essence of the process of zonal development is the progressive intensification of the zonal differentiation of the geographic envelope and the increased complexity of zonal structure. A very important role in this process belongs to the organic world. The emergence of new types of organisms was immediately reflected in the altered character of material exchange between living and non-living nature, a modification of the process of chemical element migration in the geographic envelope and the appearance of new types of soils and sedimentary rocks. Prior to the appearance of highly specialized life forms adapted to various combinations of environmental conditions, zonality on the earth was necessarily of the most primitive nature. The uniform bacteria and algae from the Proterozoic and early Paleozoic do not exhibit zonal differences; similarly, no such differences exist among the *Psilophytineae*, the first broadly spread vegetation on our planet during the Silurian and the Devonian periods. In the words of K. K. Markov, 'the primitive biosphere of the early Paleozoic could not as yet produce any notable diversity of zones and provinces, despite the presence of the necessary prerequisite, even those provided by the inorganic environment.'[10]

The marked ecological differentiation in the organic world became evident only during the second half of the Paleozoic; from then on this differentiation became a powerful factor in the transformation of the geographic envelope. It produced thick limestone and black coal horizons, it contributed to the substantial modification of the gas content in the atmosphere in the direction of that obtaining today, etc. The second half of the Paleozoic provides many other instances of the effects of geographic zonality.

Thus the organic world constitutes the most important factor in the development of the geographic envelope and, in particular, in the development of its zonal structure. Simultaneously it provides the principal index of geographic differentiation; the zonality maps of historical epochs are based primarily on palaeontological data.

The development of biological forms on the earth was stimulated in turn by the changes in environmental conditions. Radical changes in the organic world have always been associated to some degree with the 'revolutionary' stages in the tectonic life of our planet when relief was altered; the relationships between dry land and the ocean, the character of atmospheric circulation, heat conditions and moisture conditions were all substantially changed. Environmental changes of this kind gave rise to adaptive processes in organisms and assisted the development of new forms.

External cosmic and tectonic activity introduced substantial differences in the schema of geographic zones at various stages of its development, sometimes intensifying and sometimes weakening the zonal contrast and producing frequent displacement of zones. (Some geographers in fact attribute large displacements in the position of geographic zones to changes in the earth's rotational axis.) These external influences impart to the history of geographic zones a kind of rhythmic character: a sequence of epochs with relatively complex zonal differentiation and epochs with greatly reduced zonal contrasts. It is necessary to emphasize that these rhythmic fluctuations did not affect the general course of development of geographic zonality towards richer differentiation accompanied by greatly increased complexity of zonal structure. The history of zones cannot repeat itself, since it is associated with the progressive development of organic life and with irreversible changes in the inorganic components and increasingly complex atmosphere and hydrosphere.

Owing to these rhythmic influences, the development of the zonal structure of the geographic envelope proceeded, as it were, along a rising spiral. Zonal differentiation always intensified during the mountain-forming epochs, when relief became more complex and the surface area of continents increased. The latter in turn intensified the climatic differences, both zonal and provincial, associated with the different degree of continentality. During the epochs when continental relief underwent peneplanation and marine transgressions were widespread, the contrasts between zones were reduced owing to the more intensive exchange of air masses between latitudes, a more even and gentler climate, and a sharp reduction in area dominated by an arid continental climate.

According to N. M. Strakhov,[11] the following zones existed during the Lower Paleozoic: an equatorial-tropical hot and wet zone, two arid (subtropical), two humid (temperate) and two transpolar zones. During the early Devonian, when dry land occupied the maximum area, zonal contrasts must have been most pronounced. During the second half of the Devonian the continental area was considerably reduced and uniformly warm and wet climate predominated; there were virtually no arid zones.

The data concerning zonality during the Carboniferous are more reliable, since zonality during that period was clearly reflected in the organic world. One can speak with confidence of the existence of three zones: equatorial-tropical (Westphalian), characterized by arboreal club-mosses and ferns (*Calamariacaea*), and two wet temperate zones (Tungus and Gondwana Land) in which the ancient gymnosperms (naked-seed plants, e.g. the *Cordaitales* and the fern *Glossopteris*, etc.) grew, and which were inhabited by giant reptiles. During the Lower Carboniferous zonal differences were as yet indistinct and humid conditions persisted over continents. In the second half of the Carboniferous, and especially during the Permian, in connection with the Hercynian orogeny,* continental areas increased in size, relief became sharply differentiated and zonal contrasts became more acute. During the Permian the wet equatorial-tropical zone was reduced in size, and a hot, dry zone extended broadly over the northern hemisphere; a wet temperate zone and a glacial zone with continental ice were also established during this period.

In the second half of the Triassic and during the Jurassic the area of the oceans became enlarged, the surfaces of continents were strongly peneplaned, and the climate became more uniform. There are indications that there existed during those periods tropical zones with sago palms and *Bennetitales*, and temperate zones with *Ginkgoales*, certain conifers, and ferns.

The next reconstruction of the zonal system is observed towards the very end of the Jurassic, during which conditions became more arid and zonality became more highly differentiated. Similar rhythmic changes in the system of geographic zones occurred during the Cretaceous period and the Cenozoic era. But the end of the Cretaceous

* 'Orogeny' refers to mountain building with associated crustal deformation.

constitutes a particularly significant boundary in the history of land-scapes on our planet. The formation of contemporary landscape zones began during that period.

The Upper Cretaceous and the Tertiary were characterized by a substantial reconstruction of the organic world due to the widespread distribution of angiosperms (covered seed) vegetation, although the first representative of this flora seems to have appeared during the Jurassic. Owing to their better structure and incomparably greater ecological diversity by comparison with all the earlier floras, the angiosperms very soon assumed complete domination over the earth's plant cover. In direct association there developed a diverse mammal fauna as well as birds and insects; ultimately biocenoses of a con-temporary type were developed. Simultaneously the biogenic cycle of substances was intensified and grew in complexity, and contemporary soil types finally appeared.

This process was undoubtedly associated with major geological events, primarily the alpine orogenesis. The tectonic movements during the Cenozoic were characterized by exceptional intensity; they led to sharp elevational differentiation of the earth's surface, apparently greater than any such differentiation during the entire history of the geographic envelope. The rising mountain ranges and the enlargement of continents necessarily affected the general circulation of the atmosphere and intensified climatic contrasts. The development of the character of the geographic envelope during the Cenozoic was certainly not peaceful and gradual, but was characterized by a sharp variability of tectonic and climatic conditions. This variability was particularly strong in the temperate and polar latitudes.

Towards the end of the Cretaceous and during the Paleogene a warm and wet climate prevailed over the earth, and two distinct latitudinal belts have been differentiated: tropical, characterized by evergreen forest, in which the angiosperms predominated, and temperate, with leaf-shedding forests, also dominated by angiosperms. The arid zone apparently occupied a narrow belt in central Asia. During the Paleogene, leaf-shedding forest (turgay-type) grew even in the Arctic; evergreen forest (poltava-type) extended southward to the area of the present-day Ukraine.

During the Neogene, as the orogenic processes intensified, the zonal system was differentiated still further. The tropical belt became smaller and according to K. K. Markov the 'great steppe formation' took place, the temperate latitudes underwent considerable cooling, and the single unified turgay-forest zone divided into fragments. Towards the end of the Neogene the general schema of geographic zonality resembled that of the present, but during the Quaternary it was subject to further modification, mainly due to continental glaciation.*

The contemporary zonal types of landscapes, which emerged for the most part during the Cenozoic, thus turn out to be of different ages. The oldest among them are the landscapes in the equatorial zone which have changed little since the late Cretaceous or the Paleogene. Subtropical forest landscapes are also extremely old. According to E. M. Lavrenko,[12] the subtropical zone of east Asian evergreen forests existed on the same territory during the Paleogene, or at least the Oligocene. The Atlantic coast evergreen forest landscapes, corresponding to the Macronesian floristic subregion (i.e. the Azores, Madeira and the western part of the Canary Islands), are of similar age. The contemporary mediterranean zone developed as a result of the denudation and disappearance of tropical landscapes, gradual xeromorphosis,† and also because of cooling. E. M. Lavrenko determined the age of the mediterranean zonal landscape as Pliocene or late Miocene.

The temperate and polar belt landscapes underwent exceptionally strong modifications during the Neogene and the Quaternary. Their major lines of development are associated with the development of an arid climate. The arid belt, as mentioned, gradually widened. The main period during which the contemporary desert landscapes of central Asia were formed coincided with the Neogene. The steppes in southern U.S.S.R. most probably had their beginning during the Miocene, while those in the northern part of that zone began during the Pliocene although the contemporary structure of this zone had already developed during the Quaternary. The forerunners of the contemporary broad-leaved forest zones are the turgay-type forest landscapes which were widely distributed during the Paleogene throughout northern Eurasia. Today the considerably changed derivatives of turgay landscapes are represented by two isolated zonal segments of different age and structure. The far-eastern zone of mixed and broad-leaved forests is more closely associated with the landscapes of the Paleogene and preserves some of its features. In Europe, the broad-leaved forest zone (on the Russian plain this zone is regarded, in landscape classification, as a subzone of the forest-steppe zone) in a contemporary form exists only from most recent times and is considerably poorer in flora and fauna than the turgay-type landscapes.

In Siberia the broad-leaved forest zone disappeared during the

* The Tertiary is divided into the Paleogene (lower) and Neogene (upper). (Dunbar and Rodgers, 1963, p. 296)
† 'Xeromorphosis' refers to the development of plants having a thick cuticle which protects them from loss of water.

Neogene, and in its place developed the tayga which constitutes one of the youngest zonal formations. The tayga-type biocenoses apparently first appeared in mountain areas. The zonal tayga landscape on the Siberian plains, on the other hand, appeared during the Miocene or even the Pliocene, and it assumed contemporary character essentially during the Pleistocene.

The youngest zonal landscape type is the tundra. Tundra landscapes first appeared in north-eastern Siberia during the Pleistocene and spread around the pole. The tundra zone attained its present boundaries only during post-glacial times. It is important to note the development on both hemispheres of glacial zones which during the period of maximum areal size occupied approximately 40 000 000 km^2 (currently 16 000 000 km^2).

It follows that it is necessary to make a distinction between the time of the appearance of a particular zonal type (e.g. the tayga) and the time of origin and the age of a zone (e.g. the tayga, tundra, etc.) as a regional unit with regard to its contemporary aspect and its present-day boundaries. At the beginning a new zonal type occupies a small area and spreads only gradually to new territory. Considerable time elapses between the appearance of new types of landscape which form independent zones, and the development of these zones to contemporary limits. It goes without saying that these limits do not remain constant and continue to change. Individual sections of a single zone, its provinces or landscapes, are therefore of different ages. The tayga landscapes of the north-western Russian plain, for example, are younger than the Priuralye* tayga landscapes; the latter, in turn, are younger than the central-Siberian tayga.

The process of the formation of landscape zones is accompanied by movements which have a fluctuating character: a displacement of zonal boundaries to the north or the south. Such rhythmic variations in the zonal system are especially characteristic of the Quaternary. They are clearly manifest in the alternation of glacial and interglacial epochs, during which the zones were displaced by thousands of kilometres. The rhythmic displacements of zonal boundaries continued well into the post-glacial era. In particular, there was at least one period during which the tundra and the forest zone within the area of present day U.S.S.R. was displaced a considerable distance to the north, whilst in the European part of the U.S.S.R., forest reached the shores of the Barents Sea.[13]

As early as the end of the nineteenth century the hypothesis that the zones of the Russian plain are currently being displaced towards the

* An area directly south of the Urals.

south was put forward. This argument was first established for the southern boundary of the forest zone and later also for the boundary between the tundra and the tayga. L. S. Berg was one of the proponents of this view. He thought that the climate is becoming progressively wetter, causing the humid zones to displace the arid zones; as a result, the tundra is spreading over the forest zone, the forests over the forest-steppe zone, and so on.[14]

Other geographers hold a directly opposing viewpoint, namely that zones are being displaced to the north. Finally, there exists a view, held by F. N. Milkov, of the 'spatial conservation' or territorial stability of landscape zones. According to this hypothesis, zonal boundaries underwent no significant changes in the post-glacial era, and the forest-steppe zone in particular has maintained approximately identical boundaries from the Neogene right up to the present.

Nevertheless, it must be admitted that there are insufficient grounds to reject the hypothesis of zonal displacement. It is necessary to at least accept that not all zones are subject to the same displacements, nor are all these displacements strictly synchronized. It is known that zones characterizing higher latitudes are more dynamic; furthermore, it can be accepted that the occurrence of post-glacial fluctuations in the boundaries between the tundra and the tayga has been proven. In particular, the most recent studies of M. I. Neyshtadt confirm the hypothesis of L. S. Berg and others concerning the very recent displacement of the tundra and tayga, as well as the broad-leaved forest boundaries, towards the south.

It is possible, however, that these phenomena are much more weakly demonstrated in the more southerly regions of the U.S.S.R. A good deal remains obscure regarding the dynamics of Quaternary landscape zones in this area. The causes of zonal displacements also remain a contentious issue. There is no doubt that general climatic changes associated with astronomic factors play an important role; among these factors geographers include the variations of solar activity, changes in the earth's axis of rotation, and changes in the tidal force due to the mutual displacement of bodies in the earth-moon-sun system. It is highly probable that the causes of long-term variations in climate vary; correspondingly, the rhythm of climatic variations has a varying duration (e.g. there is evidence of rhythmic alternations every 20 000, 3 500, 1 800 to 1 900 years, etc.).

Rhythmic displacements of zones lead to some variations in their structure and to the emergence of new landscapes. The zones extend in part into other geological-geomorphological conditions, with resultant changes in their flora and fauna from which some elements are lost and others added. The elements that prevailed earlier in the past landscapes remain in the new zone as relics. An example is the presence of steppe relics in the vegetation cover of the tayga, a result of the tayga's displacement to the south; another example is the presence of glacial relief forms in the same zone, which is a consequence of the post-glacial extension of the tayga to the north. It follows that although the origin of landscape zones must not be confused with the displacement of zonal boundaries as fully developed natural complexes, neither should these two processes be viewed as directly opposed; they are in fact very closely related.

Azonality as a universal geographic principle

The development of our knowledge of geographic zones has been complex and occasionally contradictory. On the one hand, geographic research continued to provide proof of the Dokuchayevian zonality principle, revealing the increasing diversity of its manifestations in various natural phenomena. On the other hand, a deeper investigation of the mechanisms of geographic territorial differentiation led to the discovery of facts which frequently appeared to contradict the universality of the zonality principle. Thus it appeared that zones do not always constitute continuous belts; many are fragmented, while some exist only along the peripheries of continents (e.g. the mixed and broad-leaved forests in the temperate belt). Other zones, by contrast, gravitate to the interiors (e.g. deserts and steppes). Zonal boundaries were found to deviate more or less prominently from parallel, to the extent that in places they follow a direction approaching meridional (e.g. in central North America). Finally, a single zone may exhibit a considerable physical-geographic contrast in the meridional direction (e.g. in the tayga of the Russian plain and in eastern Siberia), while in mountainous regions latitudinal zonality seems altogether absent and in its place is found an alternation of landscape strata with altitude.

These facts prompted some geographers, especially during the early 1930s, to develop the view that the theory of latitudinal zonality was undergoing a crisis, that it needed to be thoroughly reviewed, and even that its scientific and practical significance was seriously in question. The critics of the zonal principle quoted in support of their stand such 'proof' as the lack of congruence between the boundaries of individual soil types or vegetation types with certain isotherms or isohyets (lines of identical rainfall), or the fact that there exist at the same latitude landscapes belonging to different zonal types.

It is well known that the congruence, or lack of congruence between zonal boundaries and individual partial climatic indices can in no way be regarded as a criterion of zonality. It would be even less sensible to attempt to assign a landscape to some zone solely on the basis of its geographic location. Dokuchayev and his students cautioned geographers against such primitive ideas. No member of that group ever conceived of natural zones as ideally regular belts bounded by parallel lines. Dokuchayev stated that nature is not mathematics, and that zonality is only a *principle*. The same concept was developed as early as 1898 by Dokuchayev's close collaborator N. M. Sibirtsev who emphasized that zonality is only a theory, only a generalized picture of the earth's soil aspects, and that the completeness and strict geographic sequence of zones is diversely affected by orographic, geological and climatic factors.

Lenin once wrote that 'a law, any law, is narrow, incomplete and approximate'.[15] Furthermore 'a phenomenon is richer than a law'.[16] This does not mean, of course, that laws derived by science are without value. Laws which express the causal relations underlying phenomena enable us to penetrate deeper into the essence of reality. Lenin has shown that 'the concept of a law is only one of the stages along which man discovers the unity and the connection, the interdependence and the totality of the universal process'.[17]

Each law, however, is valid only in strictly defined conditions and its effects vary in relation to strict conditions. Engels brilliantly exemplified the fact that even the so-called 'eternal laws' of physics are relative and that they constitute historical laws.[18] The fact that water is a liquid at temperatures between $0°$ and $100°$ C is a law of nature. However, it is valid only when, in addition to the existence of water and of a given temperature, a strictly defined pressure obtains. A slight difference in pressure is sufficient to entirely alter the effects of that law; deviations occur which, of course, are also subject to laws, but which appear to us as irregularities until we eventually explain their causes.*

A large number of laws operate simultaneously in nature. As a result, the operation of any one law, for example the zonality principle, must not be regarded as entirely independent of other laws which modify the concrete conditions of its application. Zonality manifests itself in different ways on land and on the surface of oceans, on plains and in mountainous areas. Geographic zones would have the form of continuous and mathematically regular belts only in a situation where the surface of the globe was absolutely uniform with respect to relief and its constituent 'material' (i.e. if no geographic principle other than zonality operated on its surface).

It follows that the various deviations and departures from latitudinal

---

* For a discussion of this idea see Ruxton (1968).

zonality are simple manifestations of the fact that zonality is not a single universal geographic principle or a single factor which could account for the complex nature of the physical-geographic differentiation of the earth's surface. In other words, all the evidence quoted at the beginning of this text which appears to contradict the principle of zonality provides no grounds on which to reject the general significance of this law for landscapes, although it does compel us to search for other causes and other mechanisms underlying geographic differentiation.

Some of these mechanisms were actually established simultaneously with latitudinal zonality. Dokuchayev himself, in addition to latitudinal zonality, established a regular alternation of physical-geographic conditions in mountainous areas with absolute altitude which he designated as *vertical zonality* and which he regarded as a basic natural principle.

It has also been known for a long time that geographic phenomena on the surface of continents vary sequentially with increasing distance from the coastline towards the interior. The relationship between this regularity and latitudinal zonality is often illustrated graphically in the so-called schematic representations of an ideal continent. Geobotanical textbooks often include such schematic representations of plant-type distribution on an ideal continent developed by Brockman-Eros and Rübel as early as 1912. A more complex schema for an ideal continent was developed in 1936 by A. V. Prozorovski.

Research into the pattern of changes in climate and in the organic world of the continents, in relation to oceanic influences, prompted V. L. Komarov to postulate the existence of *meridional zonality*. In 1921 Komarov suggested that every continent has three meridional zones: two oceanic and one continental. Later some geographers designated this longitudinal differentiation of landscapes as *sectionality*.

Other geographers, soil scientists and geobotanists, following similar ideas but in a somewhat different direction, formulated the principle of *provinciality* or *faciality*. The origins of this principle are also found in the work of Dokuchayev who showed in particular, that there exist longitudinal (provincial) differences in the soil cover of the chernozem zone. A fuller treatment of the idea of provinciality of the soil cover was given by L. I. Prasolov during 1916–22. Prasolov emphasized that soils vary not only with latitude but with longitude as well, since both the climate and other soil-forming factors, such as relief and geological structure, vary in that direction as well. Later these views were elaborated by S. S. Neustruyev, Ya. N. Afanasyev and I. P. Gerasimov.

P. N. Krylov, B. A. Keller, and other geobotanists held analogous views about the geographic regularities in the vegetation cover. Keller wrote that the effect of climate during the formation of soil-vegetation zones was superimposed over areas with a diverse geological history. This history determines differences in relief and geological structure of terrain, which in turn affect its climate. In addition the climate is subject to variations in the meridional direction. As a result, it is necessary to distinguish in each zone 'gross subdivisions in the form of regions with different geomorphological character and associated peculiarities of vegetation and soil. These regions introduce diversity and disturb to some degree the zonal continuity of change in the vegetation and soil. The reason for these phenomena is that the zone-forming effect of climate is not distributed in parallel but, as it were, interacts with geomorphological effects'.[19]

All of the mechanisms discussed so far, meridional zonality, provinciality, etc., may be grouped under a general name of *azonality*, since in the final analysis each is due to a single factor: the history of the tectonic development of our planet resulting from the action of its internal energy. The most important geographic consequences of the tectonic development of the earth are the differentiation of the global surface into continents and oceans and the emergence on continents of large structural-morphological subareas distinguished by their individual tectonic structure, material composition, macrorelief, level of altitude and sensitivity to oceanic influences.

The presence on the earth of continental elevations and oceanic depressions constitutes the clearest manifestation of azonal differentiation, which is classified as an azonality of the first order. Oceanic surfaces reflect fewer solar rays than land surfaces, and as a result the oceans absorb 10 to 20 per cent more solar heat per unit of surface than the continents. Moreover, except for the trade-wind areas, where land surfaces absorb more heat than the oceans owing to the sharp reduction of heat losses due to evaporation, the average atmospheric temperature over the oceans is higher than over the continents. Especially important factors which determine differences in the heat balance over the surface of the hydrosphere and the lithosphere are the high specific heat capacity of water and the turbulent heat exchange in the water mass. These special characteristics of the oceanic environment cause a slow heating of the oceanic surface during the summer but also its slow cooling during the winter. As a result, temperature variations are smaller over the oceans than over dry land and the return of heat to the atmosphere during the year is relatively more uniform.

Evaporation from the surface of the oceans is not limited by the available water supplies as is the case on land, and for this reason the atmosphere above the oceans contains substantially larger amounts of moisture. These differences in the properties of the underlying surface explain the principal differences between the oceanic and continental air masses.

Owing to the difference in their rate of heating, we find a continuous interaction between the continents and oceans manifested in the movement of the air masses between them. This movement adds greatly to the complexity of the general atmospheric circulation, constituting, as it were, its azonal component. The position of an area of land in the system of continental-oceanic atmospheric circulation is one of the principal factors in physical-geographic differentiation; it is reflected, via the climate, in every individual landscape feature. The further one goes from the coast towards the interior of a continent, the less frequent, as a rule, are the reappearance of the oceanic air mass, the more continental the climate, and the lower the atmospheric moisture content including the amount of rainfall and the overall moisture level. It is exactly these phenomena which constitute the causes of provinciality.

Important additional factors in the distribution of heat energy in space are oceanic currents, due in the main to the general circulation of the atmosphere but also greatly dependent on the geographic position of the continents and their configuration. In cold-current regions, such as the Californian or the Peruvian coastal areas, the ocean surface loses annually 60 (and more) kcal/cm² of heat. In the temperate latitudes the surface of the Atlantic Ocean receives from the heat flow an additional 20 (in some places more than 80) kcal/cm². The circulation of the air masses exerts, via the oceanic currents, great influence over the climate of the coastal areas of continents.

One of the main consequences of the different physical properties of land and ocean surfaces and of continental-oceanic atmospheric circulation are the differences in the temperature conditions of the interior as against the coastal areas of continents. These differences are particularly strong during the winter when the land masses become extremely cold and the seasonal maxima of atmospheric pressure occur over the continents. During such times the coastal regions subject to the intrusions of marine air masses, in the main the west coast of continents in the zone of westerly movement, are much warmer than the territories in the interiors of continents. Thus the difference between mean January temperatures in western Scandinavia along latitude 60°N. and central Yakutia, which lies along the same latitude, exceeds 40° C (Fig. 7) and the January isotherms in northern Eurasia assume a direction virtually coinciding with the meridions. During the summer, of course, the interior is hotter, though the difference is not nearly as

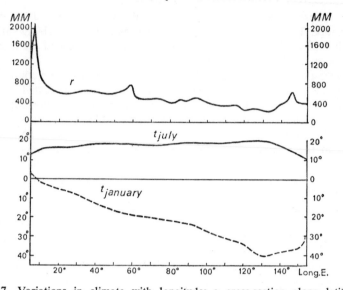

Fig. 7 Variations in climate with longitude; a cross-section along latitude 60° N.
tjanuary, mean January temperature; *tjuly*, mean July temperature; *r*, mean annual rainfall in mm.

great (e.g. the mean July temperature in central Yakutia is 6° to 7° C higher than in western Scandinavia).

The distribution of moisture over dry land depends on the movement of atmospheric masses to an even greater degree than on temperature conditions. The oceans are the principal sources of water supplies for the continents. Continents receive almost twice as much water from the oceans as from local evaporation. The major moisture mass is received from the oceans during the warm seasons when the land surface is hotter and the pressure above it is reduced, and when the barometric gradient falls away from the ocean and towards the continent and the air masses follow in that direction. The moisture content of the air in the interior regions over continents is lower the whole year round than in the coastal areas. Thus in January an air column between 0 and 7 km altitude over eastern Europe and western Siberia contains 5 to 10 mm of moisture, while in eastern Siberia it contains less than 2 to 5 mm. However, owing to the low overall moisture content in the air, the amount of winter rainfall in temperate latitudes is altogether rather low; even then, it increases greatly in the coastal regions. During the summer the amount of moisture in the atmosphere above the continents increases; in the monsoonal regions of eastern and southern Asia, where a particularly strong anomaly develops, the amount of moisture exceeds 50 mm, i.e. it is higher than on the equator. In the monsoonal regions the summer rainfall maximum is manifested with the greatest intensity.

Figure 7 illustrates the variation in total annual rainfall in the longitudinal direction along the latitude 60°N.; the peaks on the rainfall curve indicate the alpine areas of Scandinavia, the Urals and the Okhotsk Sea coastline. With regard to evaporativity, this depends to a lesser degree on longitude, although the coefficient of atmospheric moisture is subject to considerable variations. In the central Yakutsk tayga, for example, it reduces to values more characteristic of the steppe than the forest zone.

The water cycle between the oceans and the continents also markedly affects heat conditions over the continents, owing to the transport of heat in the form of potential heat of condensation which is liberated during rain precipitation. Once again, the interior regions of continents, owing to their dryness and sparse cloud cover, receive a somewhat greater total amount of solar radiation. The greatest positive anomaly in the value of total radiation is found in the trade-wind zone of continents (e.g. in northern Africa and south western Asia).

It is difficult to enumerate all the geographic consequences of longitudinal (provincial) climatic changes over continents. A marked analogy with latitudinal zonal laws is found here. In the zonal system of physical-geographic phenomena, however, the heat-and-moisture-supply processes play an equally significant role, whereas in longitudinal differentiation moisture constitutes the principal factor. The heat supply does not change very greatly in the longitudinal direction, nevertheless this factor affects the distribution of the plant cover, soils, the intensity of geochemical processes, etc.

Longitudinal changes in heat conditions are most marked in the temperate zone, where they are manifested mainly in the thermal conditions during the cold season. The severity and duration of winter increases in a direction away from the oceanic coastlines towards the continental interior, the growth period of vegetation gradually decreases, and the transition from winter to summer and from summer to winter is more rapid. As a consequence of these low temperatures there is the widespread distribution of perpetual (i.e. extending over many years) frosts in eastern Siberia.

The increasing continentality of climate in the longitudinal direction, combined with reduced moisture supplies, is clearly reflected in the soil and plant cover. The spread of broad-leaved forests is limited in the west and east only by the peripheral parts of the temperate zone, and the same is true of brown-forest soils. As for the tayga, there is a replacement of dark-coniferous forests with light-coniferous (larch) forests together with an increasing continentality of climate and a simultaneous replacement of typical podsolic soils with frozen-tayga soils.

Longitudinal physical-geographic differentiation is not uniformly developed but depends on the character of atmospheric circulation and the size, configuration and geographic position of continents. It is most fully developed in the temperate Eurasian latitudes where it is due to a vast land mass extending some 200° in longitude and subject to the west-to-east movement of the atmospheric masses. Because of the continuous supply of oceanic air masses in the west, the domination of continental air in the area of the Siberian barometric maximum and the monsoon circulation along the Pacific Ocean coast, all three meridional zones are, according to V. L. Komarov, clearly outlined.

In the trade-wind area, dominated by winds with an easterly component, conditions do not exist along the western coasts of continents for the supply of oceanic air and, as a result, deserts extend in a westerly direction almost to the oceanic coastline and the wet, near-oceanic meridional zone is entirely absent. Only along the eastern coasts of continents, owing to the monsoons, do we find a seasonal (summer) excess of moisture, which is responsible for the growth of forest vegetation, although some trees shed their leaves during the winter owing to the lack of moisture. Thus the tropical zone is characterized by a two-part longitudinal differentiation.

In the equatorial belt longitudinal physical-geographic differentiation is virtually absent owing to the slight contrast between the oceanic and the continental air masses and the weak horizontal movement; in this zone strong convection movement and abundant rainfall predominate, not only over the oceans but over the continents as well. In polar areas this differentiation is also poorly developed because of the predominance of homogeneous arctic air masses, uniform ground surface, low temperatures and excessive moisture.

As already noted, the morpho-structural heterogeneity of the land surface, i.e. the diverse character of the foundation on which landscapes develop, constitutes another factor in azonal physical-geographic differentiation. The lithological content of rocks forming the upper layers of the earth's crust, the duration of the continental development of the surface, and the character of its relief are among the most important landscape-forming influences. Their physical-geographic significance is so well known that there is no need to discuss them in detail. Rocks determine the content of migrating chemical elements and substances, essential features of hydrologic conditions of a landscape, and numerous features of its drainage pattern, as well as the content of the mineral mass in the soil.

The age of a continent, i.e. the duration of its subaerial develop-ment, often determines the fragmentation of its land surface and its drainage density. In addition, the age of a continent determines the degree to which the land surface, the structure and the vegetation species content of biocenoses, and the drainage pattern (including lake formation), etc. are developed.

Considering relief as an azonal factor, we have in mind only its broad, or *morphostructural*** features (e.g. mountain ranges and massifs, plateaux, lowland, accumulative plains, tectonic depressions, etc.) which derive their origin primarily from tectonic processes and, in the main, the most recent tectonic movements. In its morphostructural aspects, relief constitutes a powerful climate-forming factor with a wide radius of activity. The effect of rocky elevations is particularly great, not only creating specific climate conditions but also exerting a strong influence over the climates and landscapes of adjacent plains.

Depending on the orientation of the principal elements of orography, the mutual distribution of ridges and their absolute and relative heights, various changes in the general pattern of atmospheric circulation are often observed. Elevated rocks constitute barriers to the movement of air masses and accentuate climatic contrasts. High mountain ranges often prevent completely the movement of air masses, thus acting as important climatic boundaries. Ranges extending latitudinally (e.g. the Great Caucasus) prevent the passage of arctic air masses to lower latitudes, intensifying latitudinal-zonal contrasts on either side of the range. Mountain ranges extending along the meridians, especially if located along the oceanic coast, hinder the penetration of oceanic air masses to the interior of the continent and accentuate the longitudinal-climatic (provincial) differences. Such is the role of the Cordillera and the mountains in the far east, which limit substantially the wet areas over the continents bordering the Pacific Ocean.

Along the plains mountain ranges give rise to specific barrier foot and barrier shadow landscapes. The former lie along the windward side of ranges standing in the path of wet air masses, the latter are found on the lee side. The preliminary ascent of the air masses accumulated before a mountain range often begins some hundreds of

---

* The terms 'morphostructual' (or 'morphostructure') and 'morphological structure' are both used in this book. Morphostructure refers to major relief forms resulting from geological structure, while morphological structure refers to the spatial distribution and quantitative relationships obtaining among the various physical-geographic units comprising a landscape.

kilometres away; as a result, increased rainfall is observed from the windward side over a large area of the plains adjacent to a mountain range (e.g. the Colchida, the western pre-Caucasus, the near Urals etc.). On the other side of mountain ranges, by contrast, the foehn effect is observed; there is reduced cloud cover and lower rainfall, even at a considerable distance from the range. Even relatively low elevation (e.g. in the Valdai Hills or the near Volga Hills, etc.) performs a similar barrier function.

An excellent illustration of the barrier effect of mountain ranges on the landscapes of plains is provided by the Issyk-Kul basin which is surrounded by mountain ranges and extends from west to east, i.e. in the direction of the prevailing westerly air mass, over a distance of more than 200 km. (Fig. 8) Its absolute height differs very little and is on the average about 1700 m. The western part of the basin lies in the barrier shadow; here the annual rainfall is only approximately 100 mm and the major feature is a desert on grey-brown soils. The eastern part of the Issyk-Kul basin lies in the conditions of the barrier foot; here there is a gradual increase in rainfall from west to east (up to 500 mm) and the landscape changes from dry steppe on brown soils to grass steppe on chernozem-type soils.

The increased moisture supply along the piedmont plains in desert-steppe regions is often responsible for the southerly shift of the zonal boundaries and even for a characteristic inversion of latitudinal zones, i.e. a reversal of the zonal sequence. Such a situation is found in the

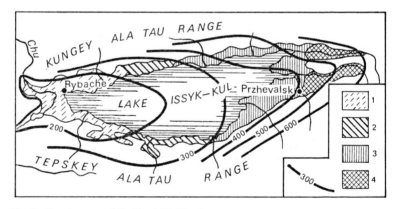

Fig. 8 Landscapes of the Issyk-Kul basin
1, desert landscapes on grey-brown soils; *2*, semidesert landscapes on light-brown soils; *3*, dry-steppe landscapes (largely cleared) on chernozem-type soils; *4*, steppe landscapes (developed in the major part) with chernozem-type soils. Isolines indicate the total annual rainfall in mm.

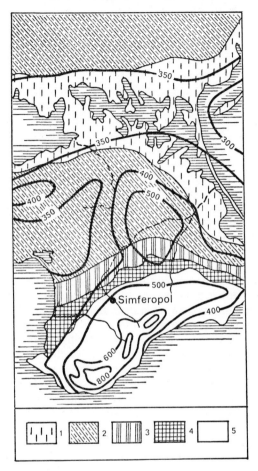

Fig. 9   Inversion of latitudinal zones on the Crimean Peninsula
1, wormwood-grass steppes on chestnut soils; *2*, fescue-stipa steppe on dark-chestnut soils and southern chernozems; *3*, mixed-grass fescue-feather steppes with chernozem-type soils; *4*, grassland piedmont steppes on carbonate cherno-zems; *5*, Crimean mountains. Isolines indicate the total annual rainfall in mm.

western pre-Caucasus and on the Crimean Peninsula. (Fig. 9) From north to south along the Crimean steppe plain, semidesert-type landscapes alternate in sequence, followed by dry-grass steppe and medium-arid, mottley-grass steppe. The cause of this lies in the increasing amount of rainfall and humidity in the direction of the mountain range. A very gradual elevation of the plain itself (from 0 to 200 m) does not play any significant role in this case.

Mountains affect lowland landscapes not only via the climate. As rocks disintegrate and material is washed away by rivers, temporary watercourses and fluvio-glacial streams, thick strata of loose deposits are formed on piedmont plains; the runoff from the mountain range feeds the lowland rivers and in the desert zone the existence of permanent watercourses and their water levels depends entirely on the rainfall precipitated in the mountains. Underground flow from the mountains also plays an essential role in the hydrological conditions of arid lowland landscapes.

A very important factor in the formation of mountainous and other elevated landscapes, apart from the uneven water supplies along the windward and lee slopes, are the different insolation conditions. These depend on the aspect of each slope as well as the angle of slope and the material making up the slope and, in particular, the absolute elevation.

Sharp rises in the absolute altitude of the land surface produce *high altitude* stratification, specifically for mountain landscapes, which constitutes a characteristic manifestation of azonality. We shall discuss this geographic mechanism separately and at this point the discussion of azonality will be summarized.

It is noted, first of all, that the two principal forms through which the azonal mechanisms of physical-geographic differentiation manifest themselves, i.e. longitudinal provinciality and morphotectonic azonality, are closely interrelated. The longitudinal-climatic boundaries usually lie along the orographic barriers (appropriate examples have been quoted). The major morphotectonic subdivisions of the earth's surface (e.g. the Russian plain, western-Siberian lowlands, central-Siberian highlands, etc.) simultaneously constitute specific climatic regions which differ from one another with respect to the degree of oceanic influence, formative conditions, movement and transformation of the atmospheric mass, and the degree of continentality, etc. It should be remembered that a single energy factor underlies both types of regularity. There are no grounds, therefore, for juxtaposing the concepts of 'azonality' and 'provinciality'.

Furthermore, it can be concluded that azonal factors affect, without exception, all geographic components; it follows that azonality and zonality constitute equivalent geographic laws. There is no area on our planet the nature of which does not depend on azonality, or which lacks certain azonal features; every section of the terrain belongs to some morphostructural element of relief, possesses some kind of geological structure, is to some degree exposed to oceanic influences, etc.

The features of the azonal differentiation of the land surface do not exhibit the kind of uniformity or sequential pattern which is found in zonal phenomena. The geographic pattern of tectonic forces is not completely understood. The existing hypotheses provide, in most cases, only very general outlines which do not explain in any detail tectonic phenomena and the distribution of geological structures in space. As a result, azonal differentiation still seems very complex and obscure.*

Whatever the causes of tectogenesis (an account of which is not the task of geography), it is important to emphasize that it does not arise from latitudinal (geographic) zonality which is associated with the distribution of solar radiation over the surface of the globe. Hence even if the effects of tectogenesis were outwardly similar to those of solar energy, they should not be confused by the use of the single designation 'zonality' which has long been applied to a strictly defined scientific phenomenon.

High-altitude landscape stratification

One of the most acute manifestations of azonal geographic differentiation is high-altitude stratification, or vertical zonality. This is produced by the sequential alternation of geographic components in mountains with height in the form of belts or *strata*, in an order essentially analogous to the sequential distribution of latitudinal geographic zones.

The cause of high-altitude stratification is a variation in heat-supply conditions with altitude. Temperature variations in the horizontal and vertical directions likewise differ considerably. In the mountains the heat balance differs from that over the lowlands. The intensity of solar radiation with altitude increases by approximately 10 per cent for every 1000 m. This is due to the reduced thickness of the atmosphere and the reduced content of water vapour and dust, so that losses in absorption and reflection of solar rays in the atmosphere are progressively reduced. At high altitudes, furthermore, the solar radiation is more uniformly distributed during the course of the year and its composition is different owing to the increase in the ultra-violet ray fraction.

However, an increase in altitude above sea-level is accompanied by intensification in the long-wave radiation from the ground surface

* One paragraph discussing the emergence of thought concerning astronomic causation of tectonic processes has been omitted.

and the rate of this intensification is higher than the rate at which solar energy is incremented. The result is a rapid drop in temperature. The vertical temperature gradient exceeds the horizontal by a factor of some hundreds. This means that in mountains, temperature falls with increases in altitude some hundreds of times faster than over plains with variations in latitude. As a result, one encounters over a few kilometres in the vertical direction changes in physical-geographic conditions equivalent to horizontal movement from the tropic to the arctic zone.

Moisture-supply conditions also vary substantially with elevation, although they do not coincide in direction or intensity with latitudinal-zonal variations. As absolute height increases, the moisture content in the air and rainfall must gradually and regularly reduce, and this is found to be the case along high-altitude plateau-like uplands. Owing, however, to the barrier function of mountain slopes discussed earlier, rainfall increases in mountainous regions up to a certain altitude beyond which it again reduces. This critical altitude is different for every mountain region and, in general, is higher in dry regions than in wet regions. Thus in the Alps maximum rainfall is precipitated at approximately 2000 m altitude, in the Great Caucasus at approximately 3000 m and in the central Asian mountains at 4000 m and higher. Since precipitation in mountainous regions is associated with the accumulation and ascent of the air mass ahead of the slopes, the windward slopes receive most of it and the lee slopes very much less. Rainfall distribution is greatly complicated by the mutual location of ridges, changes in slope gradients, dissection, and other orographic factors. As a result, the distribution of atmospheric moisture conditions in mountains is exceptionally diverse and absolute altitude plays only an indirect role.

Vertical strata differ from latitudinal zones with respect to other climatic indices as well. Quite apart from the fact that atmospheric pressure falls rapidly with increased altitude and the atmosphere becomes progressively more rarified, the circulation of the air mass becomes gradually more and more independent of the underlying surface and of seasonal temperature and pressure variations. As a rule, at altitudes above 3000 m the air is subject essentially to the laws of free-atmospheric circulation. The distribution of the cloud cover and other meteorological elements is also distinct in mountain landscapes.

The specific character of vertical strata as independent geographic complexes is not manifested only in climatic features. Specific geomorphologic processes unlike any associated with lowland landscapes (e.g. cave-ins and avalanches) are also found, the drainage pattern is quite distinct, mountain glaciers do not resemble the ice crust of the polar zone, mountain soils differ from lowland zonal soils by a thinner profile and a higher content of clastic material, etc. The distribution factors of latitudinal-zonal and vertical-stratification types of plant communities and soils are also quite different. While, for example, the northern limits of forest vegetation on plains are governed everywhere by insufficient heat, the vertical limits of forests in mountain regions are frequently governed by inadequate moisture.

Some vertical strata have no analogues among latitudinal zones (e.g. alpine meadows and high-altitude cold deserts). Latitudinal-zonal forms (e.g. trade-wind deserts) on the other hand, likewise have no analogues in mountain regions. In no way is the sequence of alternating vertical landscape strata a duplication of the latitudinal-zonal sequence, as was thought in Dokuchayev's time. Detailed investigations of vertical stratification have shown that in various mountain systems, and even in various sections of one and the same mountain region, the *structure* of the stratification (i.e. the existing strata, their number and sequence) may be subject to considerable variation. It would seem that there must exist in every mountain region lying to the south of the tayga, at some altitude or other, a mountain-tayga belt. In actual fact, the forest strata are altogether absent from many mountain regions. The same may be said about mountain-tundra and other vertical analogues of geographic zones. Furthermore, a single strata (e.g. mountain-forest or mountain-steppe) is represented in nature by a large number of variants which differ so substantially that one can hardly speak about, for example, unique mountain-forest or mountain-steppe strata. The number of known structural variants comprising the spectrum of vertical stratification greatly exceeds the number of known latitudinal zones. This is quite natural, since every strata, unlike each latitudinal zone, is essentially a local phenomenon subject to a multitude of factors which act in most diverse combinations.

The character and type of vertical stratification in a mountain range depends primarily on the type of geographic zone in which the range is located. This is clearly seen in any meridional range of sufficient length. Among such ranges are the Urals which extend over more than 2000 km from the tundra to the semidesert. (Fig. 10) In the Urals every latitudinal zone and subzone (e.g. the tundra, forest-tundra, northern tayga, etc.) corresponds with a specific system of vertical strata which has a distinct sequence, lower and upper boundaries, and certain characteristic zonal features. The mountain-tayga belt, for example, is represented in the tayga zone by dark-conifer forests, while the forest-steppe is represented by larch-pine and also birch forests.

In mountain regions with a more complex orographic structure, but

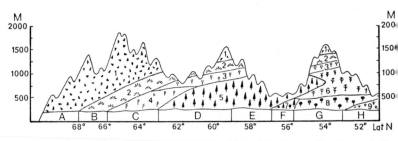

Fig. 10 Schematic representation of the high-altitude stratification of the western slopes of the Ural Mountains
1, bald hills; *2*, mountain tundra; *3*, mountain-birch stands and grasses; *4*, mountain-forest tundra and thin forest; *5*, mountain dark-coniferous tayga; *6*, mountain light-coniferous tayga; *7*, mountain subtayga (mixed forests); *8*, mountain broad-leaved forests; *9*, mountain forest-steppe. *A*, tundra; *B*, forest tundra; *C,* northern tayga; *D*, middle tayga; *E*, southern tayga; *F*, subtayga; *G*, forest steppe; *H*, steppe.

which lie along the interface of a number of landscape zones, a much more variegated picture is found, yet even in such cases it is always possible to establish some typical spectra of vertical stratification corresponding to some landscape zones. Examples of this are found in the Great Caucasus where at least six to seven vertical strata can be distinguished. Thus the lowland steppe adjacent to the northern slopes of the Great Caucasus alternates with the piedmont forest-steppe, which changes to the mountain-forest strata (first the oak, later beech and dark conifers), and above which lie in succession the strata of subalpine meadows, alpine meadows and glaciers. Along the south-western slopes of the Great Caucasus, adjacent to the Colchida (wet subtropical zone), the vertical spectrum includes the following strata: broad-leaved forests of the colchida type with evergreen undergrowth, beech forests, dark-conifer forests, subalpine meadows, alpine meadows and glaciers. Along the eastern slopes of the Great Caucasus, facing the deserts and semideserts, the strata assume a different, arid character; the forest belt contracts or is altogether absent, and where is does appear it consists of xerophytic forests on mountain cinnomonic soils.[20]

The most common manifestation of the dependence of vertical stratification on latitudinal zonality is the increasing number of vertical strata as one moves nearer to the equator. The simplest vertical stratification characterizes the glacial (Arctic and Antarctic) zones; there uniform ice and 'polar desert' strata, which constitute the extensions of the latitudinal-zonal, arctic-type landscape, are found. In the tundra it is possible to distinguish two strata: mountain tundra and glacier. In the tayga vertical stratification is even more complex: (1) the mountain-tayga strata, (2) the transitional (subalpine) strata of elfin forest and dwarf scrub, (3) mountain-tundra, (4) glaciers. The vertical stratification of the far-eastern zone of broad-leaved forest has the following sequence: (1) mixed broad-leaved and cedar forests, (2) mountain dark-conifer tayga, (3) elfin forest (comprising stone-birch *Betula ermanii*), (4) scrub and dwarf scrub, and (5) mountain tundra. It is possible to designate the typical structure of vertical stratification characterizing mountain landscapes located within a single zone as a *type* of vertical landscape stratification.

The structure of the stratification within a single zone, however, is not completely uniform. Another cause of variations are the provincial, or the longitudinal-climatic factors, principally moisture supply and the degree of continentality. The variants of zonal types of vertical stratification within certain longitudinal sections of continents are distinguished, in the main, not by the number of strata in the spectrum but by the existence of specific physical-geographic features in individual strata.

Thus all mountain landscapes lying within the tayga-type strata have a uniform system of strata, yet they reveal certain partial differences in character. In the west and the east of the Eurasian tayga the mountain-tayga strata consist of dark-conifer forest, while in the eastern Siberian mountains it consists of light-conifer (larch) forests. In the west the sparse-forest belt and the dwarf scrub are represented by depressed forms of dark-conifer species and birch forests, while in eastern Siberia they are represented by dwarf cedar scrub and in the far east by stone birch which is replaced at higher altitudes by dwarf cedar scrub and alder stands.

To each longitudinal sector there corresponds a particular set of vertical strata, which in the northern hemisphere is most distinct in lower latitudes and gradually 'contracts' to the north, while the boundaries of strata continue to descend. In the far-eastern mountains, for example, a strata of stone-birch stands, which in the broad-leaved forest zone occupies the 'third level' and does not occur below 1000 m, is found. In the tayga zone its vertical boundaries descend and it occupies the 'second level', while in the near-Pacific Ocean forests and the meadow zone (e.g. the Kamchatkan Peninsula) it constitutes the lowest strata. The strata of dark-conifer mountain tayga which constitutes the 'second level' in the far-eastern mountains constitutes the 'first level' in the tayga zone, and is altogether absent from the near-Pacific Ocean zone of forests and meadows.

Finally, the numerous local variants of the vertical stratification series arise in consequence of the orographic peculiarities of mountain regions, i.e. its direction, mutual distribution of crests and their elevation, the different aspects of individual slopes, of enclosed basins or large level areas, etc. The significance of these factors in the physical-geographic differentiation of mountains will be treated in greater detail in a later chapter, in association with the problems of structure, morphology and classification of mountain landscapes. It must be concluded that there does not exist in nature a unique system of vertical landscape strata, and that one must accept a multiplicity of vertical stratification series.

It has been demonstrated that there is only a limited correspondence between latitudinal zonality and vertical stratification, expressed mainly in certain external similarities. Vertical stratification is essentially azonal; it arises, in the final analysis, from tectonic causes and must be regarded as a special type of azonality. The phenomena of vertical stratification are not found universally, but only in areas of adequately high altitude. The concrete forms of vertical stratification depend on the latitudinal and the longitudinal position of a mountain range and on its orographic character.

It follows that latitudinal zonality and vertical stratification cannot be regarded as phenomena of the same order. If one accepts the principal differences between these two geographic mechanisms, one must reject the term 'vertical zonality' which, as it were, denotes the unity of the two mechanisms. The designation 'high altitude (vertical) stratification', or simply 'stratification', generally accepted by geo-botanists and soil scientists, expresses more accurately the uniqueness of this phenomenon.

Principal features of the
azonal physical-geographic differentiation of plains

The altitude of a territory affects not only the physical-geographic differentiation of mountains but of plains as well. The rudiments of vertical stratification have been frequently specified for certain areas of the Russian plain (e.g. the Timan and the Donyets range), the absolute altitudes of which rarely exceed 300 m. As early as the beginning of the twentieth century, G. N. Vysotski wrote that a local rise in the southern part of the Russian plain has the same effect as a transition from a dryer and warmer climatic zone to a wetter and cooler one. Vysotski also established the connection between the distribution of rainfall and hypsometry in the Ukraine, while G. I. Tanfilyov attempted to give a quantitative expression to this relationship.

Beginning with Dokuchayev, many geographers drew attention to the relationship between the distribution of various soil and vegetation types and absolute elevation. Thus according to S. S. Sobolyev, the southern boundary of the forest-steppe and the very deep Ukrainian chernozems coincides in the Ukraine with the 200 m contour, while the typical chernozems do not occur below 130 m. On the plains, however, the effect of absolute altitude on physical-geographic differentiation is different and more indirect than in mountain areas, and is frequently less significant than the other zonal factors.

In discussing vertical differentiation on the plains, it is necessary to begin by indicating the climatic differences between elevated and low-

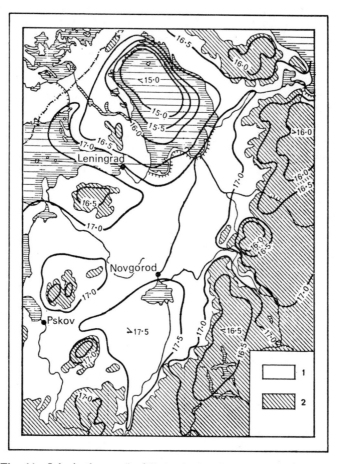

Fig. 11   July isotherms (in °C) in the north-western Russian plain
1, altitude from 0 to 100 m above sea-level; *2*, altitude above 100 m

Landscape Science

lying sections. The former are distinguished not only by a slightly cooler climate, but the vertical temperature gradient is, on the average, 0·5° to 0·6° C, while a 200 m difference in altitude should lead to a reduction in temperature by 1° C, or slightly more. (Fig. 11) In actual fact, the value of the vertical temperature gradient varies in relation to the season, exposure and steepness of the slopes, and other conditions. An example is provided by the data for the Valdai Hills and Staraya Russa meteorological stations. (Table 2) Both lie along a single latitude, the latter at a distance of just over 100 km to the east of the former, but the altitude difference between them is approximately 200 m.

It should be noted that in this case the vertical gradient for January is exaggerated, since the winter fall in temperature in the Valdai Hills with respect to Staraya Russa is produced not only by the rise in altitude but also by the increase in distance towards the east. In other words, in this case a horizontal (longitudinal) component is superimposed on the vertical gradient. If we exclude the effect of longitude, the vertical temperature gradient for January in this case will be approximately 0·4° C. Elevated localities affect the distribution of atmospheric rainfall. All the elevated areas of the Russian plain, i.e. the Vidzem, Valdai Hills, central-Russian uplands, Volga Heights, etc., receive at least 100 to 200 mm more rain than the neighbouring plains. (Fig. 12)

It is important, however, to establish the effect of climatic differences between elevated areas and lowlands on the character of landscapes. In most cases, differences in atmospheric temperature are too small to produce the effect of vertical stratification, i.e. the temperature drop would have to be two to three times greater for this effect to emerge. In many cases, elevated localities cause the displacement of zonal boundaries to the south so that the boundaries themselves assume a wavy shape, with the waves pointing towards the south along rises and to the north along the lowlands. As will be demonstrated later, however, temperature differences with altitude do not play a decisive role in these cases; moreover, the boundary zones occasionally deviate to the north along elevated localities (e.g. the northern boundary of the forest-steppe along the Volga Heights and the central-Russian uplands).

As before, the increase in rainfall with altitude over the plains is not always equivalent to the transition into a wetter zone. In the tayga, for example, where moisture is everywhere in excess, the uplands drain better than the lowlands which in the end become over-saturated and turn into swamps despite the fact that their rainfall is lower. (It can be noted from this example that actual moisture content, which

Table 2

Station	Absolute altitude (m)	Mean temperature (°C)		Temperature differences		Vertical gradient	
		January	July	January	July	January	July
Staraya Russa	24	8·3	17·5	1·4	1·3	0·74	0·68
Valdai Hills	215	9·7	16·2				

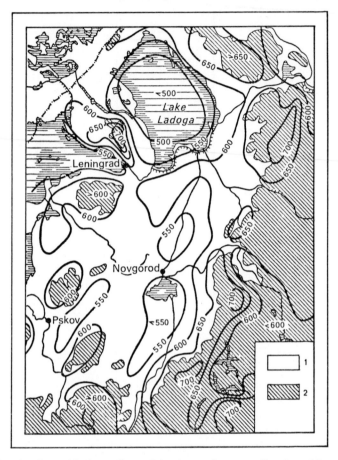

Fig. 12   Annual isohyets (in mm) in the north-western Russian plain
1, altitude between 0 and 100 m above sea-level; *2*, altitude above 100 m

depends to a large degree on drainage conditions and moisture distribution, does not coincide with atmospheric moisture.) F. N. Milkov
designated this phenomenon as inverse vertical differentiation. In
zones of inadequate atmospheric moisture the effect of increased rainfall along the uplands coincides with the effect of the temperature
drop, and the displacement of landscape boundaries to the south along
elevations is, according to F. N. Milkov, more prominent, positive
vertical differentiation.

Nevertheless, to differentiate the climatic differences associated with changes in altitude along the plains, i.e. to separate these effects from those of other azonal factors, both geomorphological and lithological, is an extremely difficult task. The effects themselves are also some-times exaggerated. Even the increase in rainfall along the upland slopes depends to a substantial degree not only on the rise of the air mass but also on the character of relief, i.e. on the roughness of the landscape. The significance of this factor is especially noticeable in coastal areas. It is well known that rainfall is lower along the actual coastal margin than at some distance from it. The reason for this is that the surface of the land, even in low-lying areas, is rougher than the surface of the water and contributes to the intensification of turbulence in the moving air masses. This pattern is clearly reflected in the system of isohyets along sea coasts and along the shores of inland water-basins (e.g. Lake Ladoga, Fig. 12). The pattern of the isotherms along the plains also depends to a large degree on the distribution of seas and lakes. Thus around Lake Ladoga the isotherms are distributed concentrically; during the summer a negative temperature anomaly (Fig. 11) is formed over the deepest section of the lake.

The following factors are generally considered to be of significance in the azonal physical-geographic differentiation of plains: the degree and character of this section of relief and the associated level of groundwater and surface drainage, as well as the lithology of the substrate, its physical-chemical properties, and the age and history of the surface. By contrast with mountain areas, the direct effect of altitude above sea-level in landscape formation is of secondary importance.

In the forest-steppe zone elevated uplands (e.g. the Volyn-Podolye, central-Russian uplands, Volga Heights, High Transvolga, etc.) are more intensely forested than the plains of, for example, the Dnepr, the Oka-Don, Low Transvolga, etc. This corresponds closely to the climatic differences between the two types of region. However, even more important in this case are certain other factors. Forest growth on uplands in the forest-steppe zone is stimulated by the dissected topography, resulting, as A. N. Krasnov has found, in better drainage and more strongly leached and lighter soils. Furthermore, the elevated areas of forest-steppe escaped the effects of Quaternary glaciations, while in the most favoured localities forest vegetation was preserved. The low-lying plains in that zone, on the other hand, were subject to glacial action as well as to the action of melting glacier waters, so that forest-steppe landscapes today are younger and more poorly drained, and their soils are frequently saline.

In a forest zone, uplands frequently result in the displacement of southern landscapes towards the north, even though it would seem that the thermal conditions there are less favourable (e.g. lower summer temperatures, reduced duration of the growth period, and lower overall temperature). Thus, for example, in the Valdai Hills the oak forests are more widespread than on the adjacent Ilmen plain, and the northern boundary of the mixed-forest zone is slightly displaced to the north in elevated areas. This is explained by the better drainage, reduced danger of snap frosts, and a more favourable substrate.

Differences in substrate often play a decisive role in the differentiation of landscape. The significance of sandy soils is well known; these assist the development of landscapes with tayga character, pine forests and sphagnum swamps, even in conditions of insufficient atmospheric moisture. Examples are provided by the landscapes of sandy alluvial plains lying in the forest-steppe zone along the Sura, Tsna and Voronezh rivers. However, where there is excess moisture the sandy substrate may have a precisely opposite effect. Sands are characterized by high permeability, are better aerated than clays and loams and dry out more quickly during spring, hence they possess properties which assist in the northerly displacement of landscapes in the tayga and tundra. Yet although the effect of these factors is observed in the majority of cases an opposite picture prevails. The reason is that sandy deposits are often spread over a water-resistant parent material (e.g. moraine, banded loams, etc.), usually along substantial depressions in relief where the sandy alluvial or fluvio-glacial material, lacustrine sands, etc. accumulate. In these areas the groundwater is close to the surface and the runoff is poor, creating conditions especially favourable for swamp formation. In addition, sands are much poorer in mineral plant nutrients than silty parent material. All these factors cause the formation on sandy alluvial, lacustrine and downwash plains of landscapes resembling more northern types.

In general, however, sand tends to smooth out the zonal contrasts between landscapes to a high degree. Zonal differences, of course, do exist on sands; they are, however, much weaker than on silt substrates. Such general features as the predominance of pine forests on poor podsolic soils, and relatively large swamp areas, characterize all low-level sandy plains from the northern tayga boundary to the forest-steppe. As the zonal hydro-thermal conditions continue to change in a southerly direction, the character of the landscapes on sandy plains also undergoes considerable changes; on desert-zone sands there are of course no pine forests or sphagnum swamps.

Another example is provided by carbonate rocks, both bedrock (e.g. limestone and dolomite) and unconsolidated material (e.g. moraine enriched by carbonate clastic material). Landscapes which develop on carbonate rock are distinguished by specific geochemical properties which are not strongly manifested in the humid, acidic forest landscapes. Because these rocks are rich in calcium, the soils are characterized by a neutral or alkaline reaction, the absorbing complex is saturated with base-exchange calcium and the accumulation of humus is intensified. The soddy-calcareous soils which develop on carbonate rocks are much more fertile than the acidic podsolic soils in typical tayga. As a result, the vegetation is richer, i.e. islands of mixed and broad-leaved forests appear among the tayga forests (e.g. the 'Silurian plateau' in the Leningrad district, and in northern Estonia, etc.). The surface and groundwater in such landscapes are distinguished by increased mineralization and hardness. The solubility of limestone leads to the widespread development of karst phenomena, e.g. sinkholes, cones and underground rivers.

It should be noted that all the factors discussed to this point, i.e. the lithological composition of rocks, the age of territory, the genesis and the dissection of the topography, as well as the hypsometric situation, are closely interrelated, and the effect of any one factor with regard to landscape formation is inseparable from the effects of the others.

At a lower hypsometric position are found the youngest aggradation plains, which were created from the accumulation of loose deposits; many were until quite recently covered with sea waters or with lacustrine or glacial-lacustrine waters. These are predominantly areas with a tendency towards lowering with regard to current tectonic movements.

At a higher hypsometric position (e.g. on the Russian plain, over 170 m on the average) are found relatively old, mainly denuded plains characterized by intensive erosion. Here the Quaternary deposits are thin, the bedrock often outcrops and is a more important element in landscape formation; eroded soils predominate and the topography sometimes approximates the topography of foothills. As mentioned, the landscapes on low-level and high-level plains also differ with respect to climatic factors.

F. N. Milkov established three levels of vertical landscape differentiation for the Russian plain: the lower level (up to 150 to 180 m), the middle level (up to 250 to 300 m) and the upper level (above 250 to 300 m).[21] The middle level may be regarded as transitional, since the two extreme types are not always sharply differentiated from one another.

In discussing the differentiating function of topography and the bedrock on plains, it is more correct to speak not about vertical differentiation but about *morphostructural* or *geomorphological* differentiation in the broad sense. The hypsometric factor plays a subordinate and essentially indirect role, while the horizontal differentiation is very much more pronounced than the vertical. The term 'morphostructural physical-geographic differentiation' of plains signifies not the effect of the sculptural details of relief (e.g. ravines, knolls and their slopes, etc.) but the landscape-forming significance of the lithological-geomorphological complexes, i.e. azonal (morphostructural) units of the terrestrial surface.

The interaction of zonal and azonal factors and the formation of landscapes

Zonal and azonal mechanisms are manifested everywhere over the earth's surface, in every geographic component and every landscape. There are no landscapes on earth which are exclusively zonal or exclusively azonal.

We still find in contemporary literature expressions such as 'azonal soils' (vegetation) or 'intrazonal soil' (vegetation).† These expressions designate soils or plant communities which are highly localized and which are subject to the effects of specific conditions (e.g. moisture saturation, bedrock, etc.), or are incompletely formed (e.g. alluvial soils). In these cases, those soils or plant communities which exist in the most typical geomorphological conditions, for example on well-drained upland areas having 'typical' soils, are treated as zonal. From this point of view it is necessary to also consider as intrazonal the pine stands in the tayga zone, since these grow on sandy soils.

There is absolutely no need to separate in this way the zonal, intrazonal and azonal soils or plant communities. G. N. Vysotski has shown convincingly that soils and plant communities developed on sands, limestones and rocky ground constitute 'the same zonal soils as the clayey and loamy soils of various types'.[22] Later B. A. Keller pointed out that the so-called zonal types of vegetation and soils are completely absent from certain regions, and yet these regions can still be assigned to a particular zone since 'the zone-determining factors leave their imprint not only on the major but also on the subordinate plant and soil types'.[23]

And, in fact, no such soils, plant communities or geocomplexes exist which are absolutely independent of the effects of the zonality law. Tidal-marsh swamp, rock, and other 'azonal' or 'intrazonal' geo-complexes turn out to be very different in different zones. They form, as it were, a parallel zonal series. It is usual practice to base the description and differentiation of zones on a hierarchical classification of soils and plant communities which have developed on upland well-drained areas. We can and should, however, compare individual zones with respect to complexes developing, for example, on sands, in depressions and in any other specific localities. These complexes differ from one another as regards moisture, geochemical conditions, soils and vegetation, as well as age and level of development, yet all belong to some particular zonal type. The formation of such geocomplexes is already associated with the mechanism of the internal differentiation of a landscape, and their investigation is the task of landscape morphology. Thus the original connotation of the terms 'azonal' and 'intrazonal', introduced towards the end of the nineteenth century by N. M. Sibirtsev to denote phenomena which are entirely zonal yet extend beyond the limits of the common conception of the zonal type, no longer has any validity.

Some geographers conceive of azonality, in its modern sense, as a secondary geographic law; azonal characteristics are viewed as purely local, and zonality is assigned a dominating role in the geographic differentiation of the earth's surface. Such a concept of azonality cannot, however, be accepted as correct, since the phenomenon itself is just as general as zonality, hence it constitutes a factor of the first order.

Azonality might also be regarded, in some sense, as the basic factor determining, through the distribution of moisture and heat, the concrete local forms through which zonal laws manifest themselves.[24] This view of zonality, however, is rather limited and does not fully express its sense. Azonal factors not only enter into the distribution of heat and moisture on the earth's surface, but they also determine the character of the solid components of the landscape and therefore the mineral composition of soils, waters and organisms. Azonal factors reflect the effects of gravitational force and, consequently, all the processes associated with gravitational energy such as the movement of water, of glaciers, etc.

It is possible to say that zonal factors constitute a specific background against which azonal phenomena manifest themselves. They strive, as it were, to level out the azonal differentiation of the continents, even though a complete smoothing out of azonal contrasts cannot be achieved since the azonal factors are also very active and continuously alter the zonal equilibrium. It follows that there are no grounds for treating either of the two factors as primary. Both comprise a kind of union of opposites in which the content of zonal and azonal factors varies in time and space, manifesting itself in different ways in natural processes and phenomena.

Atmospheric circulation alone, for example, includes both zonal and azonal components; the former comprise the ascending air currents above the equator, the trade winds, the westerly air movement, etc.; the latter comprise the monsoons and other atmospheric currents associated with land-sea interaction and with the effect of orographic barriers. Similar zonal and azonal components, however, characterize every meteorological element, e.g. the shape of the isotherms, isohyets and other climatic isolines.

An example of the combined effect of zonal and azonal factors is provided by the permafrost. The development of permafrost is strictly limited by specific zonal conditions: it is not possible to have permafrost at lower latitudes which are not subject to negative temperatures (less than 0°C.). Yet within the zones of the temperate belt the distribution of permafrost is restricted to regions with acute continental climate and extremely severe winters, both of which are azonal in origin.

Fairly complex relationships between zonal and azonal factors exist in the organic world and in soils, since the spatial differentiation of these components depends both on solar heat, moisture and the character of the mineral base.

This relationship is especially complex with regard to relief and parent material. As mentioned earlier, the processes leading to the formation of sedimentary rocks are subject to the law of zonality, but the distribution of most sedimentary rocks and of the many essentially zonal forms of relief does not correspond to contemporary zonal conditions. The reason for this is that owing to the continuous variation in the zonal structure of the earth's surface, the conditions which formed most of the sedimentary horizons of the earth's crust have disappeared a long time ago, while the rocks themselves have continued to exist over many geological epochs, constituting today the relics of past historical zonality.

In this manner the results of the zonal processes of all the preceding geological epochs, constituting the most conservative landscape components, are superimposed one over the other without necessarily reflecting the zonality of contemporary geographic processes, and are therefore regarded as azonal. Such a concept is essentially arbitrary, nevertheless it is justified by the fact that in explaining contemporary physical-geographic differentiation it is necessary to take account of the contemporary system of zonal classification, while every contradictory phenomena must be considered azonal.

Even though zonal and azonal phenomena spring from factors of dissimilar nature and are independent of one another, a substantial degree of interdependence exists between them. The history of geographic zones has clearly shown that they have developed in strict correlation with the tectonic life of the earth. During the periods of intensive mountain-formation the zonal contrasts increased sharply, and vice versa,

It has already been mentioned that the concrete forms of zonality depend on the contribution of azonal factors. In the words of S. S. Neustruyev, 'The path of zonal phenomena is different even within a single large continent, let alone two different continents, and is determined everywhere by local, climatic, geological and geomorphological factors'.[25] Among the local factors Neustruyev included those which today we designate as azonal; he understood them very broadly, or genetically, and emphasized that the 'geological factor also includes an element of the region's history'.

S. S. Neustruyev and other Russian geographers developed the idea of a series or a spectra of latitudinal zones or, in other words, systems of zonality characterizing various longitudinal segments of continents. The formation of individual landscape zone systems results from, as it were, the superposition of 'meridional zonality' (according to V. L. Komarov) over the latitudinal. Two principal types of zonal series—continental and oceanic—are distinguished, each with its own group of zones. In the continental series the desert, semidesert and steppe zones are most fully developed; continentality and relative dryness are also features of other zones including the tayga. Most typical of the oceanic series are the forest zones of various climatic belts.

The two systems are not, however, sharply divided, and there exist transitional series characterized by the wedge-like penetration of both the continental and the oceanic zonal series. As a result, in the transitional sectors are found the fullest collection of zones with a maximum number of units. An example is provided by the Russian plain, where the mixed and broad-leaved forest zones belonging to the oceanic series gradually taper off in an easterly direction until they finally disappear, while the forest steppe, steppe and semidesert zones which belong to the continental series simultaneously penetrate some distance from east to west.

An idea of the dependence of the zonal classification of land on azonal, longitudinal-climatic factors can be gained from the schematic representation of a hypothetical continent (Fig. 13) compiled by A. M. Ryabchikov and his co-workers.[26] This diagram relates land features with actual latitudinal distribution, and zonal boundaries are drawn along the mean contours from the plains of all continents, while the mountain regions are, as it were, reduced to sea-level.

So far only one effect of the interaction between land and ocean has been considered. The azonal longitudinal-climatic differentiation cuts across the latitudinal zones and which, for this reason, clearly exhibits a zonal character. But the interaction of land and ocean may also result in the horizontal exchange of air masses. In such cases, the effects of the azonal factor spread somewhat parallel to the zonal

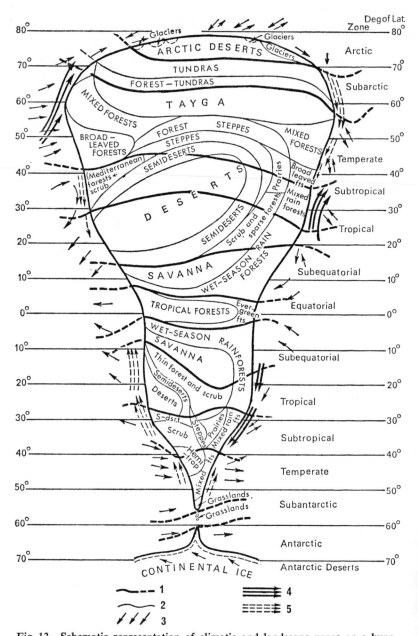

Fig. 13 Schematic representation of climatic and landscape zones on a hypothetical continent (after A. M. Ryabchikov, E. N. Lukashova, et al.)
1, climatic belt zones; *2*, landscape zone boundaries; *3*, prevailing winds; *4*, warm currents; *5*, cold currents.

effects, sometimes intensifying and sometimes weakening them, and thus leading to the broadening or narrowing of zones.

These phenomena appear most clearly in the north and the south, more strictly in the south-east of Eurasia. Along the northern limits of the continent the cooling effect of the Arctic Ocean is seen in the sharp drop of summer temperature, as shown by the high values of the latitudinal temperature gradient in the area between the coast and the southern boundary of the tundra (this gradient is some eight to ten times greater than that determined for the area between the northern and the southern boundaries of the tayga), and the distribution of the summer isotherms is almost parallel to the coastline. Such a sharp drop in summer temperatures retards the spread of forests to the north, and the southern boundary of the tundra on the whole duplicates the shape of the northern coastline of the continent. Thus although the tundra itself is unquestionably a zonal phenomenon, its southern limits, and correspondingly the northern limits of the forest zone, are largely determined by the Arctic cold. It can be assumed, therefore, that the boundary between the tundra and the tayga would lie much further to the north if the land mass extended further north.

A different picture can be found in south-eastern Asia where the latitudinal contrast between land and ocean (i.e. the Indian Ocean) intensifies the effect of the southern trade winds during the summer when the continent heats up more than the ocean. The southern trade wind extends far to the north of the equator and converts into the summer monsoon. It disturbs the continuity of the northern trade-wind belt, forces back the deserts, and is responsible for the establishment of a zone of subequatorial and tropical-monsoon forests.

In each sector the position of zonal boundaries is modified by azonal, morphostructural factors. In uniform geological-geomorphological conditions the transitions between adjacent geographic zones are very often extremely diffuse. Where, however, the earth's surface has a variegated structure, zonal boundaries assume a complex and, at the same time, a more distinct form. Thus the northern boundary of the forest-steppe zone on the Russian plain lies along the interface of morphostructural regions of two very distinct types: elevated dissected plains with loess-type carbonate soils, and low-lying sandy forest areas. The former favours the growth of broad-leaved forests and the spread of steppe elements, virtually excluding the southern penetration of tayga landscape elements. The latter, by contrast, favours a southward shift of the tayga and subtayga landscapes' swamps and conifer forests. Accordingly the boundary between the forest (subtayga) and the forest-steppe zones generally lies directly along the interface of these two types of regions. (Fig. 14)

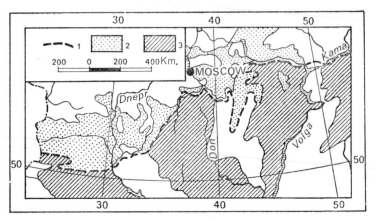

Fig. 14   Northern boundary of the forest-steppe zone on the Russian plain
1, northern boundary of the forest steppe; *2*, low-lying sandy plains (polesye); *3*, elevated dissected plains with loess and loess-like loamy soils.

In the Baltic region, owing to the widespread distribution of carbonate rocks, the northern boundary of the subtayga (the mixed-forest zone) is displaced far to the north, so that its actual position is quite at variance with the theoretical position. (Fig. 15) In fact the zonal boundary would lie much further south if the zonal-climatic criteria were utilized; but we are not justified in separating the climatic from the edaphic factors, nor could this be achieved in practice.

Contemporary landscape differentiation results from the interaction of zonal and azonal factors during the historical process of the development of the geographic envelope. This differentiation manifests itself concretely in a whole series of *regional geocomplexes* of different order, comprising a *system* of *units* of *physical-geographic regionalization* (to be discussed in chapter 4). The lowest category in this system is the *landscape*. Within a single landscape the zonal and azonal conditions remain uniform. This means that within a landscape one does not distinguish between the zonal differences in climate, vegetation and other components. No longitudinal-climatic differentiation is made; likewise no variations in the morphostructural features of topography or in the structure of the geological base, nor the associated features of all the other components is considered.

This does not mean that, in the physical-geographic sense, each landscape is absolutely homogeneous. Landscapes are frequently characterized by considerable internal diversity, and may still be subdivided into smaller parts which also constitute geocomplexes. Such differentiation, however, is no longer based on zonal or azonal

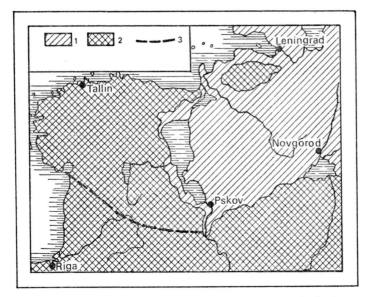

Fig. 15 Boundary between the tayga and subtayga zones on the north-western Russian plain
1, tayga landscapes; *2*, subtayga landscapes; *3*, approximate theoretical position of the tayga-subtayga boundary.

characteristics but on features of a different order which are designated as *morphological*. A detailed study of the mechanisms of internal differentiation in landscapes belongs to appropriate sections of this book, but before turning to them a few preliminary remarks will be made.

The general concept of internal (morphological) differentiation of landscapes and of the limits of physical-geographic division of a territory

At some point in a consistent analysis of the causes of physical-geographic differentiation, and along the path from the coarser distinctions to the finer details of the geographic mosaic, one begins to note that the diversity of physical-geographic conditions can no longer be explained by the effects of universal zonal and azonal factors. In a single set of zonal and azonal conditions there frequently exist, side by side, dry pine forests and high sphagnum peat-moors, forests and steppes. The alternation of geocomplexes with different microclimates, moisture conditions, soils, vegetation, and other features may take place even along a distance of some hundreds, and even tens of metres, for example along the slope of a single knoll.

Obviously in such situations we are dealing with differentiation of a different scale, associated neither with the latitudinal distribution of solar heat, nor with the continental oceanic movement of the atmosphere, nor with the morphostructural elements of the earth's crust. Rather, we are dealing with more specific or local factors effective over a small area. Such internal differentiation of a landscape may be due to extremely differing factors specific to that landscape. Such a second-order differentiation is designated as *internal* or *morphological*. Its causes are in the development of that landscape and its characteristic geographic processes are erosion, the redistribution of moisture, surface deposition and salting due to local gravitational forces, competition among plant communities and their interaction with the external environment, overgrowing of lakes, swamping of forests, salination and leaching of soils, the thawing of fossil ice, the activity of some animals, etc. It should be noted that these factors of internal differentiation act against the background of specific zonal-azonal conditions. It cannot be otherwise, since these conditions affect the overall direction of landscape development and thus also the character of all its constituent processes and its internal differentiation.

The processes of exogenous dissection of the surface and the formation of a large number of small and medium-size (sculptural) forms of topography play a particularly important role in the development of the internal physical-geographic differentiation in a landscape. The presence of such forms leads to the differentiation of localities, i.e. areas distinguished by their position in the orographic profile (e.g. various parts of a slope, sinkholes, flat sections of watersheds, peaks, etc.). Given an identical set of zonal and azonal conditions, localities will still differ as regards the character of natural drainage, water balance, microclimate, soil-forming processes and the entire complex of ecological conditions which determine the formation of biocenoses, their structure and composition.

Thus within a single locality the physical-geographic conditions have a uniform character. This means that here every geographic component is represented by its smallest territorial unit; relief by a single element, climate by a single microclimate, soil by a single phase, and vegetation by a single phytocenosis. As a group, however, all the components of a single locality constitute the simplest geocomplex, the *facia*, which cannot be further subdivided. The facia therefore represents the terminal level of the physical-geographic classification of terrain.

A small number of geographers do not accept the facia as the limit of physical-geographic classification, maintaining that no natural criteria exist for such a classification and that the terminal level should be established in each individual case exclusively on the grounds of practical expediency. From that viewpoint the facia may be sub-divided, if necessary, *ad infinitum*. Obviously the facia may be theoretically as well as physically divided into as many smaller elements as required, but it would then lose its qualitative significance as a geocomplex, splitting into single elements which are no longer objects of geographic study (e.g. grass mounds, tree trunks and stumps, turf, rock fragments, ant-hills, etc.). Thus even if further subdivision is justified in an absolute sense, it ceases to be geographically significant below the level of the facia.*

The concept of facia as the terminal geographic unit first emerged and achieved acceptance in the course of physical-geographic field surveys and landscape mapping. As a unit it was differentiated by many geographers, often independently of one another, and was often given different designations (e.g. microlandscape, elementary land-scape, etc.). The term 'facia' has proved the most viable. In later sections we shall examine in detail the characteristics of this the most simple geocomplex.

* This view is held by most Western survey organizations. A lower-order unit is defined on the basis of its practical utility and potential response to various techniques of regional resource management. See Christian and Stewart's discussion of 'site' (1968, p. 247).

3

The Study of Landscape

The definition of landscape,
and landscape as the principal physical-geographic unit

A large number of definitions of landscape exists and in the future landscapes will continue to be described and delineated with increasing accuracy. Despite minor differences, the definitions proposed by various authors are very similar. The experience of Soviet landscape science leads us to treat as basic the so-called regional concept of landscape according to which every landscape constitutes, first of all, a concrete (individual) territorial unit, and secondly, a fairly complex geographic entity consisting as a rule of many elementary physical-geographic units. Finally, a landscape is the principal physical-geographic unit and the principal object of the study of terrain. This concept of landscape was originally outlined by L. S. Berg and has undergone further development in the work of L. G. Ramyenski, S. V. Kalynesnik, N. A. Solntsev, V. B. Sochava, A. A. Grigoryev, V. N. Sukachev, V. M. Chetyrkin, and others.

According to N. A. Solntsev, the following conditions are necessary for the differentiation of an independent landscape: (1) the terrain on which the landscape is established must have a uniform geological base; (2) following the formation of this base the subsequent developmental history of the landscape throughout its area must be identical (for example, we cannot include in a single landscape two areas, one of which was and the other which was not covered by an ice-cap, or one of which was subject to marine transgression and the other which remained unaffected); (3) the climate must be identical over the entire landscape and seasonal changes must remain uniform throughout (that is, within a landscape we can have only a variation in local climate in individual urochishcha and in *microclimates* in facies). In such conditions, according to N. A. Solntsev, a strictly limited set of 'structural' relief forms (produced by a single group of exogenous factors in conditions of uniform structure), watersheds, soil types and biocenoses, develop over the surface of every landscape.

Accordingly, N. A. Solntsev defines landscape as follows: 'A landscape is a genetically uniform natural territorial complex characterized by a single geological base, a single type of relief and climate, and consisting of a set of primary and secondary urochishcha specific to that landscape dynamically associated and regularly patterned in space.'[1]

This formulation emphasizes that a landscape is a regularly structured system of simple, natural terrain complexes, i.e. it is viewed, as it were, 'from below'. A landscape, however, is simultaneously an element consisting of more complex territorial units and, in the final analysis, a component of the geographic envelope. For this reason we should approach the definition of a landscape not only 'from below' but also 'from above'. In this way the landscape is regarded as a structure resulting from the differentiation of the geographic envelope. We therefore define landscape as a genetically differentiated unit of a landscape district, zone or any other large regional unit, characterized by zonal and azonal uniformity and possessing an individual morphological structure. This formulation does not of course exclude the previous one but supplements it.

The zonal-azonal uniformity of landscape is reflected primarily in the uniformity of its base, macrorelief (i.e. the morphostructural characteristics of topography) and climate. According to Solntsev, a strictly determined set of elementary geocomplexes (facies, urochishcha), i.e. a characteristic morphological structure of a landscape, develops in such conditions. Furthermore, the zonal and azonal uniformity of landscape denotes its genetic unity, since the interrelation of contemporary zonal and azonal factors is a function of the history of its development.

A single landscape may extend over hundreds or thousands of square kilometres. In the Leningrad area a number of landscapes can be distinguished. (Fig. 16) The features of the Neva landscape are determined by its position in the southern tayga (zonal features), its relative proximity to the sea, and by the flat lowland topography on eroded, deep-lying Cambrian rocks, covered with Quaternary deposits (azonal features). The most recent stage in the history of this landscape was characterized by the emergence of post-glacial drainage basins and the accumulation of loose overburden, mainly banded loams. The result is a landscape with a relatively uniform morphological structure in which swampy forest and marshland facies predominate.

Fig. 16   Landscapes in the Leningrad district
1, north-western Ladoga; *2*, northern Ladoga; *3*, Vyborg; *4*, Lower Vuoksin;
5, central upland (Lembolov); *6*, coastal; *7*, Neva; *8*, southern Ladoga; *9*,
southern coastal; *10*, Izhora; *11*, Oredezh; *12*, Volklov.

The Izhora landscape ('Silurian plateau') has developed in similar
zonal conditions, yet genetically it differs sharply from the Neva landscape. Here the base consists of Ordovician carbonate rocks considerably elevated to form a plateau. The territory has not been
subject to intensive post-glacial flooding and the limestones are
covered by a thin layer of moraine. These limestones ensure good
drainage, the development of karst forms, the virtually complete
absence of surface drainage, the formation of soddy-calcareous soils,
and the predominance of conifer and broad-leaved forests.
    The landscapes of the northern shore of Lake Ladoga, on the other
hand, have an extremely complex morphology owing to the fact that
their base consists of crystalline rocks. Here we find a sequence of
alternating granitic ridges running from the north-west to the south-
east which differ in size and altitude. They are characterized by pine
forests of different types and rocky facies, flat longitudinal troughs
filled with lacustrine-glacial overburden, swampy forest and lakes. In

this coastal area the zonal conditions are not entirely uniform; in the south-west they have the character of southern tayga and in the north-east of the middle tayga. This is reflected in the differences in summer temperatures and the duration of the growth period, which in turn produce differences in the plant cover and certain other components. Accordingly, two independent landscapes are distinguished in this area—the north-western Ladoga (southern tayga) landscape and the northern Ladoga (middle tayga) landscape—even though their external differences are relatively indistinguishable.

To the north of the Neva plain lies the central upland landscape of the Karelian Isthmus. Here the Quaternary history is characterized by a strong accumulation of morainic material and of intra-glacial stream deposits, which elevated this landscape above the surrounding terrain. The relatively high altitude (up to 200 m above sea-level) in turn results in increased rainfall (some 200–300 mm more than along the shores of the Gulf of Finland and Lake Ladoga) and a slightly lower air temperature. The morphological structure of this landscape is determined by the contrast between the bald hill and basin formations of moraine and kame topography and by the presence of deep valleys. The plant and soil cover show typical southern tayga features.

Between the central upland landscape and the north-western Ladoga landscape we find the Lower Vuoksin lowland landscape which exhibits strong distinguishing features. During the post-glacial and late post-glacial periods this territory was subjected to intensive abrasion and accumulation. The surface horizons comprise mainly lacustrine-glacial sands with local outcrops of moraine. The bedrock lies deep under the Quaternary overburden without ever coming to the surface, but the NW.–SE. orientation of the crystalline base structure is reflected in the orientation of the larger elements of topography, i.e. fluvio-glacial troughs and lakes. The morphological structure is characterized by a large number of elevated sandy lacustrine-glacial terraces with pine forest, alternating with small moraine crests covered with spruce stands, and by the presence of wide gullies having swamps or lakes. The climate shows the influence of Lake Ladoga (i.e. reduced summer temperatures) and, in combination with the relatively poor mineral substrata, yields a soil-vegetation cover resembling that of the middle tayga, even though this landscape lies in the southern tayga zone.

It is easy to demonstrate that landscape occupies a more or less central position in the system of physical-geographic units. It was shown earlier that the causes underlying physical-geographic differentiation of the higher and lower order units are very different in nature. The derivation of regional complexes, including landscape, proceeds from causes *external* to the landscape (i.e. the zonal distribution of solar heat and the azonal differentiation of the earth's crust). Morphological differentiation on the other hand, i.e. the derivation of simple geocomplexes, is based on the action of *internal* causes, because of the regular development of the landscape itself and the interaction between its components. In this process zonality and azonality function only as underlying general conditions, providing a kind of background against which a complex set of geographic processes leading to the internal differentiation of a landscape is evolved.

It follows, *inter alia*, that the method of regionalization, i.e. the method by which we delineate and classify higher (regional) physical-geographic units, must differ from those by which we delineate and classify the lower (morphological) units. The latter are studied directly in the field by means of research stations and landscape surveys, with the obligatory compilation of landscape maps. Regionalization, on the other hand, is essentially a laboratory-type activity; field survey can only play a supplementary role. Because of this, the landscape occupies a central position.

A comparison of landscapes with regional units of a higher order (e.g. zones, subzones, etc.) shows very clearly a substantial qualitative difference: no higher order unit possesses the most important characteristic of a landscape, i.e. indivisibility with respect to both zonality and azonality. Similarly none of the morphological units (e.g. facies, urochishcha, etc.) are regarded as possessing this indivisibility. Moreover substantial qualitative differences distinguish the landscape from its morphological constituents; these provide a strong justification for the treatment of landscapes as the basic object of the physical-geographic study of terrain.

1. Elementary geocomplexes, taken in isolation, fail to reveal the typical natural characteristic of an area.* Among the mosaic of facies and urochishcha we often encounter complexes which are not typical of the given landscape, zone, district or province. In the tayga zone, for example, we can find in certain localities facies which include

* A very similar view is held by Linton in his discussion of morphological regions:
'A lithological unit, in other words, is not a physiographic region.
'An analogy from the field of chemistry may be helpful. The smallest subdivision of a substance that can exist and still retain the physical and chemical properties of that substance is the molecule. The smallest physiographic subdivision of a land mass for which description is simplified by reference to its morphological evolution is the section. Subdivision of the chemical molecule yields atoms of the constituent elements whose properties are unrelated each to the other or either to the compound whole. Subdivision of the physiographic section yields unit areas of the constituent morphological elements, closely related to the subjacent geological formations, but unrelated to each other or to the compound whole.' (Linton, 1951, p. 205)

broad-leafed forests, as well as occasional rocky facies with lichen vegetation and sparse forest which are typical of conditions far to the north. Certain urochishcha (e.g. a swamp or a gully) exist more or less independently of any association with other urochishcha and provide no basis for a comprehensive evaluation of the local natural conditions. In the words of V. B. Sochava, 'a urochishche, or a group of urochishcha, cannot constitute a basic taxonomic unit, since they do not provide us with an adequate understanding of the local structure of the geographic environment'.[2] Similar ideas are found in the work of A. A. Grigoryev,[3] who has shown that a landscape is the minimum unit which preserves all the features typical for the given zone, region, or any other regional unit larger than a landscape. None of the constituent sections of a landscape preserves these typical features in full, and therefore they cannot be regarded as 'geographic individuals'.

2. As noted by N. A. Solntsev, facies and urochishcha are not unique. Analogous facies and urochishcha frequently recur, very often within a broad and varied territory. In such simple complexes, features specifying individuality become secondary and those which specify similarity are brought to the fore, i.e. the so-called typological features. Accordingly, morphological units are regarded, as a rule, from the typological viewpoint. It follows that the geographer is not obliged to study every facia or urochishche; he can restrict himself to samples of various facia or urochishche types. By contrast, he must study each individual landscape and he has no right to sample; this of course does not exclude the need for establishing any existing similarities between landscapes and their typology.

3. A landscape differs from elementary geocomplexes in its longer history and greater stability to external influences, including that of man. Even in strongly transformed landscapes the basic zonal and azonal factors continue to act and the character of many components is preserved. Facies, on the other hand, change very swiftly and often fundamentally, retaining no traces of their original structure.

4. Elementary geocomplexes cannot be regarded as independent material systems, since they cannot exist apart from one another. Any facia or urochishche can exist only in association with one or more strictly defined facia or urochishche. A facia on the lower part of the slope exists only in association with that on the upper part. Gully urochishcha could not form if there had been no watershed urochishcha. The morphological units of a landscape always comprise conjoined or associated systems. Within such systems the diverse facies and urochishcha are mutually associated into a whole by the processes of runoff and the migration of chemical elements, by the local atmospheric circulation, and by the processes of settlement and

periodic migration of organisms. They are unified by the general effect
of zonal and azonal conditions, i.e. their common history of development.

It is impossible to gain an understanding of an individual facia or
urochishche if we analyse it in isolation from its neighbours and insist
on isolating it from the genetically homogeneous system of which it is
a constituent. Furthermore, it is impossible to understand, within the
framework of a single elementary geocomplex, the processes occurring
in landscape and their dynamics. Only a full set of associated morphological units can provide the basis for geographic investigation. In
other words, the principal objective of research in geography must be
a territory within the boundaries of which the entire set of elementary
geocomplexes, facies and urochishcha, is manifested in typical association, i.e. a landscape in its regional sense.

Landscapes are also not isolated from one another, since in nature
nothing really exists in isolation, but the bonds between landscapes are
much more limited and essentially indirect (e.g. via atmospheric circulation). For the development of a landscape, the interaction with
its neighbours is not as decisive as the interaction among the urochishcha or the facies. In other words, a landscape is a much more
autonomous natural system than either of the others. In investigating
landscapes we can attempt a comprehensive study of geographic
interrelations and process which cannot be done with respect to
simple geocomplexes.

The concept of landscape as a basic unit of physical-geographic
terrain differentiation is supported by the practical approach to landscape as an object of economic activity. V. B. Sochava and V. M.
Chetyrkin have shown that a landscape constitutes precisely the kind
of natural territorial complex for which one could propose uniform
economic development. In determining the economic potential of an
area we obviously cannot begin at the level of facies or urochishcha,
since these units give an incomplete account of the local natural
conditions in which economic activity is carried out. We can gain such
an idea only by considering the landscape holistically. Groupings of
various ugodya* within a landscape create conditions for the development of various activities, especially farming. The individuality and
the associative existence of the individual components of a landscape
provide a natural-scientific basis for economic and land-reclamation
decisions.

In contemporary geographic literature the term 'landscape' is frequently used in a more general sense, i.e. to designate the geocomplex

---

* Agricultural land.

as a whole, independent of its size or taxonomic status. Such a concept of landscape was common during the early stages of development of landscape science, prior to the establishment of the various geocomplexes and before the qualitative differences between them were determined. At that time even colonies of mushrooms in a forest or the mounds of earth at the entrance to squirrel-burrows were regarded as landscapes. Even today some geographers designate the entire Russian plain as a landscape, as well as a single gully or any other fragment of the earth's crust. Some geographers attempted to provide a theoretical backing for this concept, basing their arguments on the boundlessness of physical-geographic classification and the absence of any qualitative differences between geocomplexes of different orders. This concept has been strongly disputed by many geographers.

Without touching on the theoretical aspects of this problem, it is obvious that from the terminological viewpoint alone the application of the term 'landscape' to such diverse geographic objects as the Russian plain and a single gully, or the central-Siberian tablelands and a spruce stand, is irrational. It not only creates difficulties as far as descriptions are concerned, but it can easily lead to confusion, especially since there already are used in landscape science two synonomous 'general' designations—a natural territorial complex and the geographic complex (geocomplex)—and we do not need a third synonym.

The academician A. A. Grigoryev noted appropriately that since all geographic units (geocomplexes) are constituents of the geographic envelope, they must necessarily share certain properties in common; but this does not permit the designation of each separate unit as a landscape, as to do so is to radically alter its meaning in the manner opposite to that prevailing today.

Finally, there exists yet another, though not very widespread, typological concept of a landscape. It emerged during the 1930s but did not gain much support. According to this concept a landscape is a type or species of geocomplex, i.e. it is not a piece of terrain but a set of typical properties characteristic of some of its sections, independent of their rank and distributional nature. It is well known that the concept of type can arise only if concrete objects exist. In other words, the idea of concrete phenomena is primary and the idea of 'type' is derived. When we speak of a 'type', we invariably ask, A type of what? We can speak of types of facies, types of urochishcha, types of landscapes, etc. We cannot speak of type or species in general. Moreover, in the typological concept of landscape, the idea of type exists, as it were, apart from any association with concrete geocomplexes. This concept eliminates whatever differences there are between geocomplexes of different orders. It does not reflect contemporary knowledge regarding the morphological structure of landscapes which has gained widespread acceptance among geographers.

The regional concept of a landscape combines both the individual and the typological approach to the study of geocomplexes and presupposes their obligatory systematization with respect to typological aspects. It takes as given the existence of geocomplexes of different kinds and order, and the priority of concrete 'individual' geographic units, and regards the ideas of species and type as the result of subsequent theoretical generalization. The regional concept therefore avoids bias characteristic of the typological view. Designating this concept as regional signifies no more than the idea of a landscape as a concrete regional complex and the principal unit of physical-geographic regionalization.

The content and structure of a landscape

A landscape includes the mutually associated components of the solid crust, the water resources and the atmospheric envelope, together with its biocenoses and soils. All these form the substantive components of a landscape. Furthermore, climate and relief (topography) are usually listed as components of a landscape. The former relates to a complex of the properties and processes of the atmospheric envelope and the latter to the inseparable and specific character of the solid crust.

We should note that all these components enter not only into the structure of landscapes but also into the structure of any other geocomplex, from facies to the geographic envelope. The facia, for example, also comprises the solid substrata, some hydrospheric and atmospheric elements, vegetation and living forms. Landscape zones and other geocomplexes of a higher order share the same components. For example, the geomorphological and soil characteristics of a facia, landscape or zone cannot be expressed in the same way; each component manifests itself differently in geocomplexes of different orders. Accordingly, we need to refine the idea of landscape components; it is obviously insufficient to speak broadly of 'relief', 'soils', etc.

Each geographic component distinguishes specific categories of terrain differentiation, and it is important to establish which of these categories correspond or conform to a landscape. It has already been noted that the smallest subunits of geographic components (e.g. elements of relief, microclimate, phytocenosis, etc.) constitute the components of facies. Landscapes, on the other hand, include more complex 'component units'.

Every definition of landscape stresses its uniform geological base. This means primarily the homogeneity of lithological structure and uniform conditions of occurrence of surface rock strata. These characteristics are in turn associated with the structure of a folded base. This association, however, is often indirect, and the effect of the folded base often fails to manifest itself, especially in the case of ancient plateaux where the folded base is buried under a thick layer of sedimentary rocks, as for example over large areas of the Russian platform and the western-Siberian plateau. Much more pertinent are the most recent tectonic movements which determine the major forms of relief.

It is not necessary for the geological base of a landscape to consist of a single type of rock. The concept of the lithological structure of the base best conveys the idea of geological formation as a genetic rock complex which forms in certain specific structural-facial conditions.*

On lowland plains with undisturbed sedimentary strata on a deep-lying folded base, almost uniform rocks often extend over very large areas. One example is the north-western Russian plain where a sequence of alternating bands of sedimentary strata of different ages and lithological structure are found. (Fig. 17)

In the mountains folded bedrock underlying a single landscape may comprise a complex of thin alternating rock strata of different origins and structure but associated within an independent structural element of the earth's crust (e.g. a synclinal or anticlinal structure, a sequence of small folds, etc.). Often the landscape consists of intrusive outcrops or lavas; such landscapes are found, for example, in the trap-

---

* 'Facial' used in this context refers to stratigraphic conditions of part of a geologic *system*. The term is derived from the word 'facies', employed originally by the geologist Gressly in 1838. The following quotation from Gressly is taken from Dunbar and Rodgers, *Principles of Stratigraphy*:

'I have succeeded, in this manner, in recognizing, in the horizontal dimension of each formation, diverse well-characterized modifications, which present constant peculiarities in their petrographic constitution as well as in the paleontologic characters of the assemblage of their fossils, and which are governed by appropriate and hardly variable laws.

'And at once there are two principal facts which characterize everywhere the assemblages of modifications that I call facies, or aspects of formations: one is that a given petrographic aspect of any formation necessarily implies, wherever it occurs, the same paleontologic assemblage; the other is that a given paleontologic assemblage rigorously excludes fossil genera and species frequent in other facies.

'I think that the modifications, whether petrographic or paleontologic, shown by a formation in its horizontal extent, are produced by the different locations and other circumstances that so powerfully influence, nowadays too, the different genera and species of organized beings which people the Ocean and the seas of the present.' (Dunbar and Rodgers, 1963, p. 136)

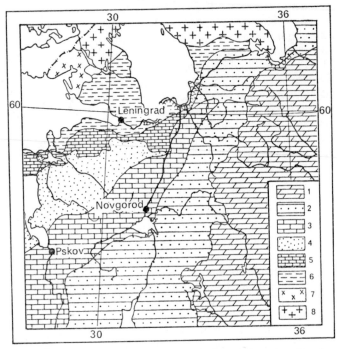

Fig. 17   Geological base of the landscapes of the north-western Russian plain
1, Lower and Middle Carboniferous limestones, dolomites and clays; *2*, Upper
Devonian multicoloured deposits (sands, sandstones, clays with marl and limestone bands); *3*, Upper Devonian limestones, marls and clays; *4*, Middle
Devonian red sands; *5*, Ordovician limestones; *6*, Cambrian sandstones, sands
and clays; *7*, Upper Proterozoic crystalline rocks (rapakivi-granites); *8*, Archaean
acidic crystalline rocks (granites, granodiorites, etc.).

cuts* of central Siberia and in the lava plateaux of the Armenian
uplands.

Plains, too, often include sequences of lithologically and strati-
graphically different rocks. These are found wherever a deep valley
exposes strata. In such cases we find outcrops of younger rocks in
drainage divides, with older rocks along the slopes, and still older
rocks in the valleys. (Fig. 18) Despite this, we are justified in saying
that the geological base of such landscapes has a uniform structure,
since the alternation of rock strata is due not to geological processes

---

* 'Trap-cuts' is a general term for bedrock composed of dolerites, diabase,
diabasic porphyrites and basalt. Trap-cuts develop as a set of flat platforms
resembling a staircase.

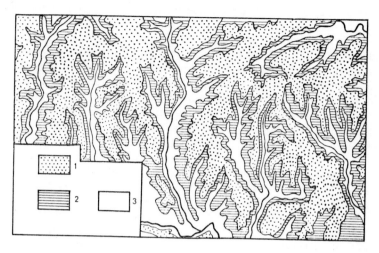

Fig. 18   Geological base of the southern part of the central-Russian high-lands
1, Oligocene sandy-clayey deposits; *2*, Eocene sands and sandstones; *3*, Upper Cretaceous deposits (blackboard chalk).

but to the internal dissection of landscape, which should be taken into account when landscapes are classified into morphological units.

Directly associated with the geological base is surface relief or topography. When relief is considered as a landscape component, it should be remembered that it includes the major features of the continental surface or the oceanic deep. In other words, there exist different categories of relief. Geographers use freely such expressions as 'macro-relief', 'mesorelief' and 'microrelief', but these are essentially inconsistent and there is not as yet an acceptable classification of relief forms despite the many attempts to derive such a classification.*

We are primarily interested in the classification of relief forms into morphostructural and morphosculptural, as mentioned in chapter 2. The first includes major continental features of azonal origin, i.e. those produced by tectonic movements and geological structure (e.g. upthrusts of various kinds, plateaux, mountain ranges, etc.). The second, i.e. the sculptural forms, comprise the detail against the

* James favours the use of the terms 'microchore', 'mesochore' and 'macro-chore', meaning small, medium or large regions (or areas). (1934, p. 85) It is his opinion that mesochore should refer to regional studies; this suggestion does not seem to be widely accepted by Western geographers.

background of major features, due mainly to the action of the exogenous denudational and accumulative agents, and are essentially
zonal in character (e.g. various erosional, glacial, wind-borne, and
karst forms, etc.). These sculptural relief forms are closely associated
with structural forms, since the dominating effect of the exogenous
geomorphological factors (accumulation or denudation) depends on
the underlying features produced by tectonic uplifts and downthrusts.
In regions with uplift tendencies, the complex of denudational, sculptural forms predominates; large depressions on the earth's surface,
however, are affected by different types of accumulative forms of
relief.

A genetically uniform morphostructural element of the earth's
surface, together with its set of sculptural forms, may be regarded as
a *geomorphological complex*. Such a complex is commensurate with
landscape. It has a uniform geological base and is subject to the same
exogenous geomorphological processes; in it are unified the ideas of
geological structure and landscape relief. A geomorphological complex
therefore constitutes a 'hard component' of landscape, or a segment
of the lithosphere corresponding to landscape. Crystalline shields
with a complex of glacial denudational and fluvio-glacial accumulative
forms, structural plateaux composed of limestones and dolomites and
capped with glacial and karst forms, and intermontane tectonic
depressions filled with proluvium, alluvium and other deposits, are
examples of geomorphological complexes. Similar (recurring) geomorphological complexes comprise a single type of relief. Occasionally a geomorphological complex does not coincide with a single landscape. We often find in territories constituting a single geomorphological complex certain zonal or longitudinal-climatic variations
responsible for the emergence of two or more independent landscapes
(e.g. the north-western Ladoga landscapes mentioned in the previous
chapter).

The hydrosphere is represented in continental landscapes by
extremely diverse forms which could hardly be grouped under a
common designation. V. I. Vyernadski, who regarded natural waters
as special kinds of minerals, classified them with regard to their
physical state (gaseous, liquid, solid), concentration of salts (fresh
waters, salt and briny waters), character of course localities (surface
and ground water), sources of origin (lake, swamp, river, etc.) and the
chemical composition of dissolved materials. In terms of Vyernadski's classification, there exist some 500 such water 'minerals' on
earth, although more detailed studies indicate that the number could
reach between 1000 and 1500. The diversity of natural waters is, of
course, directly associated with the character of the landscape. Every

landscape has a regular system of water reservoirs with characteristic dynamics, chemistry, heat conditions, etc.

The relationship between the landscape and climate was successfully resolved by S. P. Khromov,[4] who has shown that the differentiation of climate into categories of territorial dimensions arises directly from the differentiation of the geocomplexes themselves into different taxonomic units, since climate constitutes only one component of the geocomplex. As his principal unit Khromov takes the climate of a landscape, which he designates simply as *climate*. The climate of a urochishche, which represents a characteristic variation of landscape climate, is designated as *local climate*, and the climate of a facia as *microclimate*. Khromov also introduces the idea of *macroclimate*, which specifies the complex of climatic conditions over a given geographic region or zone (i.e. the dominating categories of regional physical-geographic classification). The observations of each meteorological station characterize primarily the local climate, i.e. the climate of the specific urochishche in which the station operates. To gain an idea of the climate of a landscape it is necessary, in general, to work with data from several stations distributed over the most typical urochishcha.

The organic world of a landscape consists of a more or less variegated complex of biocenoses. By contrast with the facies, a landscape is not characterized by any one plant community, any one association or formation, or any other category of plant classification. A single landscape may include associations belonging to different types of vegetation (e.g. almost every landscape in the tayga zone includes forest, swamp and grasslands, and occasionally even tundra and other types of vegetation). Similarly, a specific plant formation or association may extend over many landscapes. It follows that landscapes are not characterized by any 'specific' associations or formations, etc., but by regular groupings of various plant associations (and biocenoses), which form a characteristic (so-called topo-ecological) series with a specific distribution over the facies and urochishcha. These topo-ecological series are the basis on which the geobotanical classification of regions is constructed. In practice, this means that independent geobotanical regions are territorially commensurate with landscapes.

A similar relationship exists between landscapes and soil regions. It is difficult to find landscapes with only a single type of soil. Different soils frequently alternate over a small area, each associated with a single facia. Accordingly, every landscape embraces a regular territorial complex of soil types, subtypes and varieties which correspond to a single soil region.

It is necessary to distinguish between the components of a land-scape and its elements. The elements of a landscape are the parts of individual components (e.g. plant and animal species, rock types, atmospheric gases, etc.). Whereas each component of a landscape forms the subject of investigation for some geographic discipline, the elements of landscape are not geographic objects at all but belong to other natural sciences.

As well as the material components, it is necessary to recognize *energy components* of a landscape. The most important of these is solar radiation. The energy of tectonic and volcanic processes mani-fests itself directly in tectonically active regions of the globe. The indirect effect of the earth's internal energy, however, is just as constant and universal in character as the action of solar radiation energy. The processes of mechanical displacement of materials over the earth's surface play an important role in the development of a landscape. The movement of material masses over the surface of the lithosphere is rendered possible by topography, which affects the specific distribution of areas and locations with different gravitational force potentials. But the topographic base (tectonic relief) results from the action of the internal energy of the earth. Accordingly, when materials are moved mechanically over the earth's surface, the action of the gravitational force 'masks' the energy of tectonic processes, which in various past geological epochs was lost in overcoming just that gravitational force.

Gravitational forces manifest themselves in the troposphere and the hydrosphere when differences arise in the density of the atmospheric and water masses. These differences are due to non-uniform heating, i.e. the uneven amounts of absorbed solar heat. Thus a proportion of the solar energy is converted in the atmospheric and hydrologic environment into the kinetic energy which moves the air and water masses.

The interaction among the individual units of a landscape deter-mines its *structure*, i.e. the internal organization of objects and phenomena within this complex material system. Among the struc-tural units we include not only the components discussed earlier but also the morphological units (facies, urochishcha, etc.). S. V. Kalyes-nik also includes among the structural features of landscape its seasonal rhythm. According to Kalyesnik, the structural landscape comprises the following three types of features: the character of inter-action between individual components, the character of associations among the morphological units, the major features of seasonal rhythm manifest in the change of aspects.[5]

A landscape has a considerably more complex structure than the simple geographic units, e.g. the facia. In facies every component has a uniform structure; the relationships between such components are more direct. In landscapes, on the other hand, we are concerned not only with the interactions of various components of uniform segments of terrain (i.e. facies) but also the interaction between the segments themselves which manifests itself in very diverse forms (e.g. the distribution of heat and moisture, the migration of chemical elements, etc.). The interaction between the morphological units of a landscape determines its *morphological structure*.

We shall devote subsequent sections to the problems of morphological structure and the seasonal dynamics of a landscape; at this point we shall examine the contribution of individual components to the development of the structure of a landscape. In general, the role of these components should be clear from the study of general geography* and little additional comment is required. Primarily, it should be noted that each component plays a specific role in a landscape, and one could hardly compare this role against some scale which treats one component as more important than another.

Geographers have often tried to divide landscape components into major and subordinate, usually assigning to the former, relief and geological structure or climate or both. And in fact these components of a geocomplex seem to play a major role, not only because of their age on the time-scale of the earth's history, but also because they constitute, as it were, the initial links in the chain of geographic interactions. We can imagine landscapes virtually empty of life and water, but we cannot imagine landscapes without hard ground surface or without climate.

The climate and the geomorphological complex are the first components subject to the direct action of external zonal and azonal factors. These factors affect other components only through climate and through the solid base. Accordingly they play the most important role in the spatial differentiation of physical-geographic conditions and in the formation of landscape boundaries. In this, and only in this, sense can we treat them as major components.

The lithosphere forms the base of a landscape. The material of the lithosphere enters into the organisms, the soils, the water and the atmosphere of a landscape. The hard base is the landscape's most stable and conservative component. It preserves very efficiently, in the form of different rocks and types of relief, the effects of the physical-geographic conditions of past geological time, and the legacy

---

* The study of landforms, soils, vegetation, etc. in general, without reference to their association in a particular place. See chapter 1, p. 3.

of long extinct landscapes. The great diversity of material content and of forms of the external surface of the lithosphere is the principal cause of the variety of landscape distribution and the most important factor in morphological, intra-landscape differentiation.

The specific role of the atmosphere, as opposed to that of the lithosphere, arises from the exceptional mobility of the atmospheric environment. The movement of air masses is in itself a powerful mechanical factor affecting the earth's surface, transporting and depositing mineral particles, seeds of plants, etc., and contributing to the formation of relief. More important, however, is the transportation of heat and moisture within the air masses. The properties of the ground layer of the atmosphere are determined not only by the conditions prevalent over the given landscape but by the effects of other, sometimes very distant, landscapes. In this way landscapes interact through atmospheric circulation.

Landscapes are formed not only by climate but predominantly by the actual content of the atmosphere; atmospheric oxygen is the main source of oxidation reactions on earth, carbon-dioxide is the principal material in the structure of organic matter, while water vapour, a major factor in the heat conditions on the earth's surface, is a source of rainfall and likewise an important factor in the regulation of heat conditions on the earth's surface.

Water plays a substantial role in mechanical weathering in a landscape and is an important geochemical factor, providing an environment in which most chemical reactions occur. Going through its cycle, water passes from one physical state into another, penetrating every component of the landscape, entering into their structure and altering their properties by promoting exchange among them. Most chemical elements are transported by it; runoff is a factor in the distribution of mineral matter among landscapes and among the morphological units of a landscape.

The inorganic components constitute, on the whole, a major category with respect to the organic components; organisms build their bodies from elements of the atmospheric, hydrospheric and the geological envelopes. Thus the vegetation and the animal life, as well as the soil, are often treated as minor or subordinate components. We must bear in mind, however, that the present-day content and structure of all the three inorganic geospheres result from a prolonged period of interaction between living and non-living nature, in which the inorganic components play a passive role while the organic play an active role. Vyernadski wrote that living matter is the most constantly acting and the most powerful chemical force on earth; it is capable of transforming radiation energy and of conserving it together with the organic material which is continuously accumulated in the earth's crust.

Because of this continuous material exchange, organisms absorb into the biogenic cycle the inorganic units of the geographic envelope. Since this cycle has been taking place continuously over billions of years, it has fundamentally transformed the initial atmosphere, hydrosphere and lithosphere. According to Vyernadski, oxygen, nitrogen and carbon-dioxide in the atmosphere are of biogenic origin, i.e. the present-day atmosphere has been created by organic life. Organisms played an important role in determining the gas and ionic content of natural waters and their chemical properties. Most of the moisture evaporating from the earth's surface passes through vegetation which thereby automatically assumes an important role in the water cycle. Organic life participated directly or indirectly in the formation of all sedimentary rocks.

The action of organic life on rocks, supplemented by solar energy, moisture and air, produces a separate, wholly active component of the geocomplex, soil, within which the interaction of living and non-living nature finds its strongest manifestation. Soil-forming processes, in turn, react to moisture conditions, the development of biocenoses and the formation of sedimentary rocks.

The role of biocenoses in the morphological differentiation of a landscape is considerable; they form characteristic local climates and microclimates, and they tend to overgrow small bodies of water leading to heat-accumulation, and to affect runoff, erosion, weathering and other processes.

In the geographic envelope the most stable and immobile material is the lithospheric matter, the particles of which are extremely coherent. Owing to the constant water circulation in the lithosphere, the penetration of oxygen and carbon-dioxide, and especially the action of organic life, engage the lithosphere in the water cycle, transform and enrich it. The stability and permanence of the solid base has prompted some geographers to treat the lithosphere as the major component of a landscape,[6] even though logically the conclusion should be completely the opposite. The geological base of a landscape can preserve fundamental features over many millions of years; the landscapes themselves during this time (i.e. those extending over the same base) will have altered very many times. The type of relief may also retain its essential features with a change of landscape. There is a good reason why relief is often called the conservative or disassociative component. It is obvious that a component which changes more slowly than the others cannot play a dominating role in the development of a landscape. The leading role in the developmental processes is always claimed by the most dynamic or progressive components. Such a component in the geographic envelope is organic material which represents qualitatively the highest and most active form of matter.

Thus the interaction among geographic components assumes complex and diverse forms. To gain some knowledge of these processes we must analyse landscape development. The division of components into major and subordinate is therefore relative; it is also static in character. It makes sense for some isolated time span in the process of landscape development but becomes irrelevant when we examine the interaction of geographic components on a historical scale.

Landscape boundaries

Landscapes are separated from one another by natural boundaries which can be very diverse in character. The transition from one landscape to another signifies a change of structure which involves both a change in components and a change in morphological structure. Wherever landscape structures alter prominently, landscape boundaries are distinct. And, vice versa, wherever the structure alters gradually, which is usually the case, landscape boundaries are more or less diffuse.

Landscape boundaries are due to the action of zonal and azonal factors. This action, however, is transformed in the landscape and manifests itself mainly via the geomorphological complex and the climate. For this reason the distribution of landscape over an area is directly determined by many factors, one of which may have a dominating role, e.g. a sharp rise in altitude above sea-level, a change in the type of bedrock, zonal or azonal variations in climate, etc. (Fig. 19) It would be incorrect to postulate that a single dominating factor determines the boundaries of all landscapes. Even a single landscape may have a boundary the origins of which differ from one place to another.*

Since all components in a landscape closely interact, a variation in one must somehow affect the others. A change in landscape over an area must be reflected in the climate, in soils, in vegetation, etc. Accordingly, landscape boundaries constitute, as it were, complex multiple boundaries, often of a very intricate nature. Only rarely do

* This is quite often the case rather than the exception. An obvious Australian example may be found in the Grampians mountain area where Sibley mapped the *Grampians Ranges* land system, a mountainous cuesta dipping west and striking north–south, juxtaposed to at least eight other land systems and varieties of land units within each system. In some cases a boundary is indicated by significant changes in landform, in others by soil or vegetation. (Sibley, 1967, map of land systems)

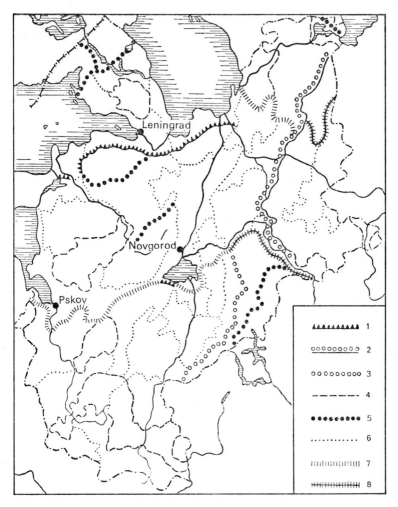

Fig. 19   Types of landscape boundaries on the north-western Russian plain
1, distinct orographic boundaries—high escarpments associated with an alternation in bedrock; *2*, as above, but complicated by the accumulation or morainic
material and subsequent abrasive action; these appear as broad slopes; *3*, as
above (*1*), but with a deep-lying bedrock base (the orographic boundary is paramount to the geological boundary); *4*, orographic boundaries dividing morainic
uplands from low-lying plains; *5*, boundaries created by bedrock alternations,
poorly evident in relief; *6*, poorly manifested transitions due to gradual variations of absolute altitude, drainage conditions, and the lithologic and structural
nature of Quaternary deposits; *7*, diffuse zonal-climatic boundaries; *8*, zonal
boundaries derived from orographic boundaries.

Landscape Science

landscapes have distinct borders. For this to occur, all the components must alter sharply and simultaneously, and such situations are exceptional.

Every geographic component has its specific pattern, which includes the individual character of its transition from one area to another. Certain characteristic boundaries (e.g. climatic, hydrological, zoogeographic, etc.) are naturally diffuse; others (e.g. geobotanical, soil, etc.) may be either fairly sharp or diffuse, while geological and geomorphological boundaries are on the whole quite distinct.

The degree to which boundaries manifest themselves in components is directly associated with their mobility or variability over time. Although the development of individual components is mutually interrelated both in time and space, it proceeds at uneven rates. Every component has a kind of momentum with respect to underlying causes; some react fairly rapidly to changes in conditions, others do so more slowly.

General climatic changes on a global scale, conditions of solar activity, or other external factors, are first reflected in macroclimate and lead to the displacement of climatic zones. Climatic changes are soon inevitably reflected in hydrological, soil and biogeographic processes. Biogeographic boundaries follow climatic boundaries though with some delay but soil changes are much slower than biogeographic changes. Relief reacts even more slowly to changes in zonal boundaries. Man's influence very rapidly alters the vegetation, the wild life, and hydrological conditions. Landscape changes over time depend not only on external factors but on their own development as well. We shall not discuss here, however, the causes governing landscape development; it is sufficient to note that they do vary with time and their boundaries shift accordingly. Owing to the non-uniform development of individual components, landscape boundaries are somewhat diffuse.

Frederick Engels wrote that 'Hard and fast lines are incompatible with the theory of evolution.[7] For a stage in the outlook on nature,' Engels continued, 'where all differences become merged in intermediate stages, and all opposites are bridged by intermediate links, the old metaphysical method of thought no longer suffices.'[8] The old metaphysical method required 'sharp' and 'stable' boundaries. Engels wrote that 'For everyday use, for the small change of science, the metaphysical categories retain their validity.'[9] In practice the geographer too is obliged to follow the metaphysical approach when, for example, he depicts on a map a boundary between two geocomplexes by means of a line. This is a conventional representation; it does not signify a departure from reality; it merely reflects the imperfection of the method available to us for the representation of reality.

Landscape boundaries are most distinct when they are due to factors which also vary sharply in space. Azonal factors, owing to their permanence, create the most stable landscape boundaries. Relatively clear landscape boundaries are associated with the alternation of surface rocks of different composition (e.g. carbonate and non-carbonate, sandy and clayey, etc.) even in cases where such alternations are poorly reflected in topography.

Such azonal boundaries are often emphasized by the hydrographic network. Many rivers flow along the boundaries of different morphostructural units, their one bank hugging steep valley sides and the other marginal to low plains and broad terraces, e.g. the Volga, the Oka, the northern Donets over much of its length, the Dnepr in parts, and many other rivers.

Since azonal factors exhibit more frequent and contrasting variations over space, most landscape boundaries are of azonal origin. Many boundaries, however, are due to changes in zonal conditions. In cases where zonal and azonal boundaries coincide, the transition from one landscape to another is usually relatively sharp. If, on the other hand, significant geomorphological contrasts exist in the transition area between zones or subzones (e.g. in border areas of the large plains in western Siberia and on the Russian plain) and changes in landscape structure are primarily determined by the gradual change in climate, landscape boundaries are most indistinct.

It is true, of course, that even the most distinct landscape boundary possesses some width and can never be represented accurately by a solid line. The following example shows how complicated the nature of landscape boundaries can be. (Fig. 20) The western boundary of the Valdai Hills is defined by a sharp change in the nature of bedrock. Stable limestone and carbonaceous dolomites create a high scarp ('the carbonaceous glint') facing west. But the orographic boundary does not coincide with the geological, since it was formed by the accumulation of morainic material in front of the scarp. The climatic boundary, in turn, precedes the orographic, since already at some distance ahead of the rise, the air masses begin their ascent and precipitation increases. In this case the orographic boundary is the one which stands out most clearly and is therefore treated as a landscape boundary, even though such a choice is essentially arbitrary.

Boundaries formed by rivers or by sea and lake coastlines are variable since the level of the water surface continuously fluctuates, so that the landscape boundary has in fact the shape of a transitional band between the upper and lower water levels.

It was noted at the beginning of this chapter that a boundary between two landscapes is determined by changes in morphological

126 *Landscape Science*

Fig. 20   Schematic diagram of the relationships between geological, orographic
and climatic boundaries
1, permanent bedrock; *2*, loose bedrock; *3*, Quaternary deposits (moraine);
4, direction of air currents.

structure, i.e. changes which occur in the content of morphological
units, their quantitative relationship, mutual distribution, etc. The
boundaries of morphological components are usually more distinct
than the boundaries of the landscape itself. The landscape boundary
cannot be compiled from the boundaries of individual urochishcha,
however, since the latter are not always easily assigned to either of
their neighbouring landscapes; certain types of morphological units
are common to both landscapes, others gradually wedge in as one
landscape gives way to another. Thus even if it is accepted that the
boundaries of the urochishcha are always distinct, it does not follow
that the boundary between two landscapes will also be clearly defined.

Along the boundary between the north-western Ladoga and the
Lower Vuoksin landscapes (Fig. 16) we find, over a distance of a few
kilometres, the gradual thinning out of the selga urochishche characteristic of the former. This is accompanied by a change in the
character of the selgas: their size is reduced, they become lower and
more dense. Although every selga urochishche is relatively well
defined, it is impossible to draw a sharp line between the two landscapes; their boundary is a transitional band in which the typical urochishcha of one landscape gradually give way to the typical urochishcha of the other.

This example confirms Engel's point that 'there is no leap in nature,
precisely because nature is composed entirely of leaps.'[10] The transition from one landscape to another in space and time constitutes a
gradual qualitative change made up of numerous 'leaps' from one
facia or urochishche to another.

So far, the discussion has concentrated on the boundaries dividing one landscape from another, i.e. horizontal boundaries. Landscapes, however, are three-dimensional objects extending upwards and downwards from the surface of the lithosphere. It is possible, therefore, to pose the problem of the *vertical limits* of a landscape. This problem has not been adequately investigated and existing hypotheses are often contradictory.

According to one such hypothesis, a landscape extends from the upper to the lower limits of the geographic envelope, thus embracing its entire width. The division of the troposphere into components does not, however, conform with that of the lithosphere. The differentiation of the layers of the lithosphere is subject to azonal factors, while zonal mechanisms are predominantly active in the troposphere. The structures of the earth's crust are distinguished by relative inertness and stability, as against the exceptional mobility of the atmospheric mass. For this reason the morphostructural units of the crust do not necessarily correspond to the specific masses of the troposphere, which means that the geographic envelope cannot be divided from the top down into parts which would be characterized by uniform tectogenesis, atmospheric and other processes. Such uniformity exists only along the narrow interface at ground level, where landscapes are formed.

Yu. P. Byallovich put forward the idea of the characteristic 'stratification' of the geographic envelope. Every taxonomic unit of terrain classification corresponds to a certain layer ('chorohorizon') of the geographic envelope. According to Byallovich, the higher the rank of a terrain unit, the greater the thickness of its chorohorizon. Although this idea appears quite logical, it does not entirely reflect reality.

To begin with, the existing systems for the vertical stratification of the troposphere (i.e. the ground layer, friction layer and free troposphere) and the lithosphere (i.e. the pedosphere or the soil layer, the weathering crust and the stratisphere) are not brought into terrain (horizontal) classification. S. P. Khromov treated horizontal and vertical subdivision of climates as completely independent. The ideas of 'microclimate', 'local climate', 'climate' and 'macroclimate' refer to horizontal climatic subdivision and, according to Khromov, are completely unrelated to vertical climatic subdivision. For example, the climatic nature of the ground layer of the troposphere may be determined both by numerous microclimates, by local climates, the climates as such (i.e. the climates of individual landscapes) and even by the macroclimate. It should not be assumed, therefore, that the microclimate constitutes the climate of the ground layer and the macroclimate that of the free atmosphere.

Many atmospheric phenomena (e.g. cloud cover and rainfall), independent of their altitude, characterize equally zones, provinces and landscapes. The purely theoretical proposition that the enhanced taxonomic significance of a geocomplex elevates its upper boundary in the atmosphere, therefore makes very little practical sense.

The upper limits of a landscape are essentially indeterminate simply because the properties of the air over any surface area are determined not only by the physical-geographic conditions of that area but also by the effects of other often very distant landscapes. In addition, even if such boundaries specifying the direct effect of a single landscape on atmospheric processes for a given moment of time could be defined, the mobility of the air mass would shift the boundaries from one moment to the next.

As for the lithosphere, any unit of landscape classification must have a lower boundary, but the position of that boundary depends not so much on the rank of the geocomplex as on its origin. Such units as major landforms (e.g. the Russian plain, the Urals, central Siberia, etc.) reach with their 'roots' deep into the lithosphere, right down to the lower limits of the geographic envelope, since major landforms, determined by azonal factors, belong to the major geostructural categories of the earth's crust. Zonal units, on the other hand (e.g. zones and subzones) cannot reach as deeply into the lithosphere, since their existence is associated with the effect of the energy of solar radiation. According to A. A. Grigoryev, zones and subzones belong to the upper stage* of the geographic envelope. Furthermore, on continents the lower limit of the 'upper stage' lies at a depth of only 15 to 30 m. Thus the zonal and azonal geographic boundaries lie along different depth levels.

As already noted, there exists a comparatively narrow interface which embraces the top layer of the lithosphere—namely the core of weathering—and the ground layer of the atmosphere where both the major sources of energy, and therefore the zonal and azonal phenomena, are equally strongly manifested and to some degree inter-dependent. Only within this layer can the components of the geographic envelope, which are characterized by uniform zonal and azonal factors, exogenous and endogenous processes (i.e. landscapes) be isolated. The contact layer of the geographic envelope (i.e. the interface) corresponds to V. I. Vyernadski's 'level of life' on the earth and to E. M. Lavrenko's 'phytogeosphere'.[11] We are dealing in this case with the surface of land, but analogous layers exist over the surface of the oceans and over the ocean bottoms.

* A relatively thin layer of the geographic envelope including the surface of the earth.

The lower limit of this interface on continents is defined by the depth at which are observed the different effects of the material and energy components of a landscape. These are the effects of the characteristic geographic processes: the transformation of solar energy, the water cycle, weathering, the geochemical activity of organisms, etc. All processes at the interface are characterized by a seasonal rhythm. It is known that annual temperature variations are manifested in the tropics to a depth of 5 to 10 m, in the temperate belt to 15 to 20 m, and at high latitudes to 20 to 30 m. An important geochemical feature of the interface is the predominance of oxidation processes associated with the generation of free oxygen. The limits of penetration of free oxygen into the crust generally coincide with the upper level of ground water. The oxidation zone has a maximum thickness of approximately 60 m, and in some strongly fissured rocks up to 300 m. The thickness of the core of weathering varies from a few metres to some tens of metres, and occasionally to 100 m and more.

Organisms, i.e. certain types of bacteria, penetrate to a depth of some 3 km into the lithosphere. Organic life is most active, however, in the top layer of the earth's crust, that is in the core of weathering and in the soil, which are several tens of centimetres and sometimes a few metres thick. In this layer the major mass of the underground root systems of plants, micro-organisms, insects, earthworms, etc. is concentrated. Some rodents penetrate to a depth of 5 to 6 m, and some worms to 8 m. Roots of plants may penetrate into the bedrock to a depth of some tens of metres. Fish have been found in groundwater at a depth of over 100 m. Thus the lower limits of the most important components of landscape are very similar, not more than the order of a few tens of metres. The boundaries of a landscape penetrate the crust to the same depth. Those of facies and urochishcha virtually coincide.

Landscapes also include the ground layer of the atmosphere. Landscape boundaries on this level, however, are not very definite. We must obviously include in landscape the entire level of air through which the vegetation penetrates (i.e. up to 15 to 20 m in the forest zone of the temperate belt, 40 to 50 m in tropical forest, and slightly more in certain localities). But the effect of vegetation on the movement of air is much higher above the tree-crowns; the pollens and spores of plants are transported above this level and many species of insects spend a major part of their lives there. E. M. Lavrenko considers that the biogeocenosis includes an atmospheric layer stretching up to 10 to 50 m above the vegetation. As already noted, included among the properties of a landscape are certain climatic phenomena observed at an even greater height. According to S. P. Khromov, landscape climates also extend to certain altitudes in the free atmosphere. It follows that landscapes have no strictly defined boundaries in the atmosphere.

The same is true of the lower boundary in the lithosphere. Rocks unaffected by weathering and soil formation provide the base for a landscape and enter gradually into the cycle of matter. In the landscape lies the connection between the atmosphere and the lithosphere and the continuity between contemporary and historical landscape-forming factors. This confirms once again S. V. Kalyesnik's view that every landscape is inescapably bound up in all respects with the geographic envelope.[12]

The lower layers of the troposphere (above the ground layer) and the upper strata of the bedrock (below the weathering crust) may be regarded as the *external* stages of a landscape, within which its boundaries are gradually erased. The properties of the upper layers of the troposphere and of the bedrock depend only to a limited degree on the landscape, and their differentiation does not coincide with that of the geocomplexes in the interface layer. However, in studying the origin of a given landscape, its structure and other properties, it is difficult not to consider the underlying levels of the crust and the higher levels of the atmosphere, i.e. we are forced to take into account the entire thickness of the geographic envelope.

Landscape morphology

The term 'landscape morphology' is used to designate that part of landscape science which studies the mechanism of internal terrain classification of a landscape and the mutual interrelations and interactions among its morphological units.

A comprehensive morphological description of a landscape requires the solution of the following problems: (1) the number of categories (levels) in morphological classification and their taxonomic relations; (2) the typology of morphological units (separately for each category, i.e. for facies, urochishcha, etc.) and their characterization; (3) the spatial relationships among morphological units, i.e. their mutual distribution and their relationships with respect to area; (4) the mutual relationships (associations) among the morphological units of landscapes, and the connections between the major and subordinate units; (5) the origin of the morphological structure and its dynamics.

Virtually every landscape can be consistently divided into morphological units of a different order. The principal categories of morphological classification of landscapes, facies and urochishcha, were established in 1938 by L. G. Ramyenski and later elaborated in considerable detail by N. A. Solntsev. The concepts of facies and urochishcha have gained universal acceptance and are regarded generally as well founded.

A facia is the elementary physical-geographic unit comprising a uniform locality and habitat and a single biocenosis. Facies are formed within the limits of a single type of mesorelief, with uniform bedrock, and uniform hydrological conditions, microclimate and soil.

Urochishcha are associations of facies on individual convex or concave sections of relief, or on interfluvial plateau sections with uniform substrata, and are characterized by the common direction of water courses, the uniform transport of solid materials, and the migration of chemical elements.*

The great diversity of landscape morphology does not always allow facies and urochishcha to be organized into a simple two-unit system. Morphologically complex landscapes exhibit up to five or six levels of internal subdivision. Thus it is possible to distinguish between the facies and the urochishcha an intermediate level—a group of facies

---

* 'Facies' and 'urochischa' are sets of terrain units recognized at different scales, as the above brief description implies. Modern survey organizations working more or less independently in various parts of the world, e.g. in Canada, England, Australia, Russia and Africa, designate similar 'lower' units. There seem to be at least two basic, relatively small units. The first represents virtually uniform landform, soil, and vegetation conditions. This is the Russian 'facia', the 'site' of the CSIRO Division of Land Research, the 'component' of the Soil Conservation Authority of Victoria, the 'land element' of the Military Engineering Experimental Establishment, the 'physiographic site' of the Ontario Department of Lands and Forests, and so on. The second represents associations of the first; in other words, relatively small, uniform areas are grouped on the basis of their similarity and/or contiguity. These larger units include the Russian 'urochshche', the 'land unit' of the CSIRO Division of Land Research and the Soil Conservation Authority of Victoria, the 'landtype component' of the Ontario Department of Lands and Forests, the 'land facet' of the Military Engineering Experimental Establishment, and so on. A review of Russian literature concerning landscape science from *Soviet Geography: review and translation* indicates that this terminology, especially of the 'lower' units, is fairly widely accepted in Russia. It is quite likely, however, that in most Western countries where more than one survey organization is involved in general-purpose, terrain classification, terrain unit terminology varies from organization to organization. Difficulties arising from comparison of reports done by different organizations are somewhat ameliorated by referring to operational definitions usually outlined in the introductory section of each report.

Intra- and inter-national terminological non-conformity is undoubtedly characteristic of any emerging scientific discipline. Although the advantages of standardizing terminology are obvious, it is doubtful if any universal schema will be accepted before internationally applicable principles for classification and regionalization of terrain as well as a sound methodology are firmly established and accepted generally. Such international concurrence is unlikely in the near future, if the lack of agreement, dating from the early 1700s, concerning traditional regionalization in geography is any indication. A final stumbling-block to the establishment of a universal terminology is the vagueness surrounding definitions of the purpose of survey and included variables.

or a sub-urochishche—and between the urochishche and the land-scape a complex urochishche or a complex of urochishcha—a site.* But these additional morphological units often have only a local significance and may be relevant for only a particular group of land-scapes. In attempting to create a more complex, multi-level system of morphological units appropriate for all landscapes, considerable difficulties are inevitably encountered. Such a system would not be very useful, for it would entirely deprive landscapes of their unique-ness. To ensure a contrasting analysis of landscapes with respect to their morphology, a minimum number of obligatory categories is sufficient. Experience has shown that such obligatory categories are the facia—constituting the natural terminal unit of morphological classification—and the urochishche—a well-defined unit of shape, usually clearly distinguishable in aerial photographs, and present in virtually every landscape. The other morphological units should be differentiated as the need arises.

Recent literature uses the designations 'sub-urochishche' and 'site' for the intermediate morphological categories, and many authors are inclined to treat them as obligatory. According to a definition proposed by the staff of the Landscape Science Laboratory of the Moscow State University, the sub-urochishche is a 'natural territorial complex comprising a group of facies genetically and dynamically associated due to a common location on a single element of mesorelief with uniform aspect'.[13] Thus a sub-urochishche constitutes a part of the urochishche, which in turn embraces groups of facies which replace one another along an individual slope. For example, if an entire valley corresponds to a single urochishche, each of the two opposing slopes as well as its floor can be regarded as separate sub-urochishche. The differentiation of sub-urochishcha has a practical significance in large-scale landscape surveys (at scales of 1:10000 to 1:25000) which renders impossible the differentiation of individual facies. The sub-urochishcha are clearly defined only against a strongly dissected topography: there is little point in trying to differentiate them over broad, upland areas.

The relationships between urochishcha, sub-urochishcha and facies are illustrated by the schematic map compiled by the Landscape Science Laboratory of the Moscow State University. (Fig. 21)

The term 'site' has recently gained fairly widespread acceptance,

* The term 'site' (Bourne, 1931), as applied to terrain classification by Western geographers, is used generally to refer to relatively small and uniform areas of land; a site is analogous to a facia when used in this way. We have here translated the Russian word, *myestnost*, as 'site', and use this term to indicate a complex of urochishcha or a complex urochishche.

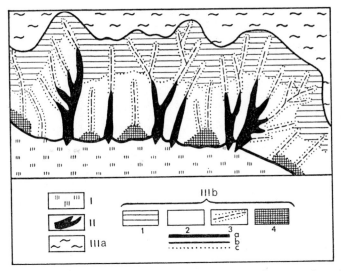

Fig. 21   Fragment of a schematic landscape map showing a section of the north-east of the central-Russian highlands (after N. A. Solntsev, et al.)
Urochishcha: *I*, dry steppe ravine with grass and mottley-grass meadows on soddy soils; *II*, spreading gullies in loessic loams, soil cover underdeveloped; *III*, undulating elevated plain with loessic and gravelly-loam cover and grey-forest and chernozem-type soils (largely under cultivation), with isolated oak forests. *Sub-urochishcha*: *IIIa*, upland area on loessic loams, poorly leached heavy-loam chernozems (under cultivation); *IIIb*, southern exposure ravine slope on leached, podsolized and dark-grey forest soils, washed-off to a varying degree (under cultivation). *Facies*: *1*, upper, gentle part of slope (1.5° to 2° gradient) with poorly washed-off medium-leached heavy-loam chernozems; *2*, middle part of slope (3° to 5° gradient) with medium washed-off medium-leached medium-loam chernozems; *3*, washed-out rill with dark-grey strongly washed-out medium-loam forest soils; *4*, deluvial foothill areas with washed-in podsolized heavy-loam chernozems. *a*, urochishcha boundaries; *b*, sub-urochishcha boundaries; *c*, facia boundaries.

but its meaning is not consistent. Quite often it is used loosely to mean the 'geocomplex' or the 'natural terrain complex'. Some geographers, for instance F. N. Milkov, define 'site-type' as a group of areas with different genesis, zonal and azonal physical-geographic characteristics and united only by their common location (e.g. the 'upland type of site', 'the terrace-above-the-flood-plain type of site', 'the riparian type of site', etc.). Finally, some geographers, K. I. Gerenchuk, N. A. Solntsev and others, designated by the term 'site' the largest morphological part of the landscape, and that which constitutes an individual variant of a group of urochishcha characteristic for a given landscape.[14]

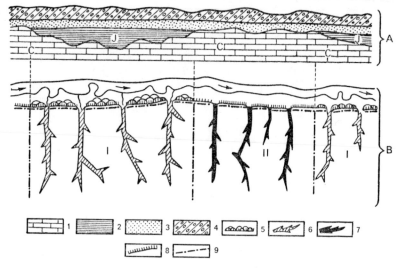

Fig. 22 Schema illustrating the conditions for the differentiation of a site as a morphological unit of a landscape (after N. A. Solntsev)

A, geological section: 1, Carboniferous limestones; *2*, Jurassic clays; *3*, sub-morainic fluvio-glacial sands; *4*, moraine.

B, section of a river valley with adjacent sites: I, site consisting of urochishcha of wet gullies, complicated by landslides, on Jurassic clays; *II*, site with dry-gully urochishcha cutting through fissured limestones; *5*, landslides; *6*, wet gullies with landslides; *7*, dry gullies; *8*, river-bank escarpment; *9*, site boundaries.

N. A. Solntsev associates sites with variations in the geological-geomorphological base of a given landscape. He illustrates this by means of a schema shown in figure 22.[15] Solntsev also presents the following example: in monticulate-morainic landscapes are found, alongside the sections in which the urochishcha of large morainic hills alternate regularly with those of large basins, other sections where tiny knolls and small basins alternate. These two variants of a single landscape serve as examples of sites.

In many cases sites correspond to large forms of mesorelief. One example is provided by the landscape of the Pobuzhe Ranges northwest of Lvov, described by K. I. Gerenchuk, in which two types of sites, range and inter-range, alternate.[16]

These ranges are approximately 25 km long, 1 to 5 km wide and 25 to 35 m high. They are characterized by the following groups of urochishcha: (1) upland, on the flat summits of the ranges, with podsolized chernozems; (2) hollows along the summits, with thin soils along the slopes and chernozem-meadow gley soils in depressions; (3) steep slopes, with dark-grey and grey soils; (4) gully and ravine urochishcha.

The inter-range valleys consist of flat, swampy sites, $0\cdot5$ to $2\cdot0$ km wide, with the following urochishcha: (1) a marginal strip on shallow cretaceous marls, topped with humus-calcareous (rendzina) with fine loamy soil; (2) sections with occasionally excessive moisture and thick soddy-meadow gley soils; (3) swampy sections of valleys with peaty-gley soils and sedge-grass vegetation; (4) peat bogs.

Upland landscapes with well-developed interfluves, terraces and flood plains, each with its own set of urochishcha, may be regarded as individual sites. In some cases sites correspond to areas of a single landscape which differ not with respect to the qualitative structure of the urochishcha, but only with respect to quantitative ratios (e.g. a frontal morainic apron with areas of swamp urochishcha varying in size).

Finally, fragments of alien landscapes within a given landscape may also be regarded as individual sites. These are typical of the youngest continental glaciation landscapes with their complex dissection and contrasting alternation of genetically diverse sections. Thus among monticulate-morainic landscapes, sections of frontal apron and lacustrine-glacial plains are often found; among the morainic or lacustrine-glacial plains we often find sections with kame hill-and-basin complexes, etc. In such cases Leningrad geographers speak about different urochishcha complexes; a landscape, however, is also a complex of urochishcha and in such circumstances the term 'site' would be more meaningful.

The concept of 'site' as a morphological unit cannot yet be regarded as firmly established. Many units designated as sites have only local significance and cannot be easily compared with sites differentiated in other landscapes. It is possible that it will prove more convenient in future to distinguish a number of independent categories of landscape classification in place of the present-day site. Some sites would be more correctly viewed as *complex* urochishcha (these will be discussed in due course); sometimes complexes designated as sites in fact constitute independent landscapes.

The notion of 'site' as applied to the morphological classification of mountain landscapes needs a more precise definition. At the present time the term 'site' is assigned in mountain landscapes both to sections of high-altitude strata and to units of landscape classification associated with orography or lithology (e.g. monoclinal, medium-elevation ridges, relics of an ancient peneplain, areas with saline karst, etc.), and confusion naturally results.

An important problem is the *typology* of facies, urochishcha and other units of morphological classification. Without a typology, it is impossible to comprehend the enormous diversity of morphological units or compile a landscape map. This problem is currently being studied and a good deal of attention is being given to the typology of urochishcha, since the latter are often subjects of landscape mapping. We shall return to this problem later.

Another problem currently requiring development is the typology of morphological structures of landscapes, i.e. the grouping of landscapes according to the type of morphological structure, taking into account the character of the mutual distribution of morphological units and their qualitative and quantitative relationships. So far there have been only sporadic attempts to distinguish the types of morphological structure of landscapes, and only with respect to a few small areas. It is therefore too early to speak about a comprehensive classification of landscapes from the point of view of morphological structure, and we shall restrict ourselves to a few typical examples.

Many landscapes exhibit a regular, rhythmic alternation of the same morphological elements oriented in a single direction. This group includes the Pobuzhe Ranges landscape discussed earlier, the selga-hollow landscapes of the Karelian tayga, the crest-and-trough landscapes of the western Siberian forest-steppe, desert landscapes with sandy ridges, etc.

Very often, however, morphological units do not recur in a landscape but replace one another in some specific direction, e.g. the landscapes of piedmont sloping plains. K. I. Gerenchuk quotes the following example: the landscape of the Tisen plain (Transcarpathia) consists of a series of urochishcha of various types, changing gradually in the position of the slope from the most elevated, dry urochishcha with oak and white beech forests on soddy, podsolized soils to low-level, undrained urochishcha with swampy grasslands, peat bogs and alder stands on silty-swamp and peat-bog soils. A separate variant of this type is represented by landscapes with steppe-like morphological structure, characterizing terraced, alluvial, lacustrine or marine plains.

Very often morphological landscape units are not distributed with any particular consistency. We can speak in such cases about different variants of the mottled or mosaic morphological structure. One example is provided by the hill-morainic tayga landscapes with randomly scattered moraine hillock urochishcha, covered with dark-conifer forests, swampy troughs and lakes. (Fig. 23) A characteristic type of morphological structure with its so-called micro-complexity,

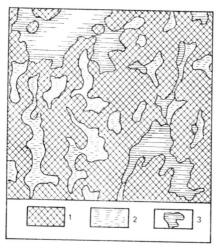

Fig. 23 Morphological structure of the hill-moraine landscape in the southern tayga
1, urochishcha of morainic hills with spruce forest; *2*, urochishcha of swampy depressions; *3*, lakes.

found in landscapes of semidesert plains, is produced by small patches of different facies.

Another type of morphological structure characteristic of many steppe-plain landscapes has upland urochishcha providing a background for the sporadic distribution of subordinate urochishcha on sink holes, minor depressions, etc., with mesophytic plant communities on solonetzic soils.

The same landscape may combine the features of morphological structures of different types. Also large structural components of the first rank (e.g. sites) consist of units of second rank (urochishcha) which produce a type of secondary morphological outline. The facies, in turn, form a characteristic pattern within the urochishcha (e.g. in many urochishcha—convex, ridge-and-mochezhina swamps, solonchak depressions, etc.—there is a typically concentric-band distribution of facies). It should be stressed that landscapes of one morphological type, if their similarity is only external, may prove very different genetically.

The rhythmic alternation of urochishcha of two different types (ridge-hollow, etc.) may be due to the structural-morphological properties of rocks, the effect of erosion, wind, etc. The phenomenon of micro-complexity results from the influence of microrelief on the

distribution of moisture and salts and on microclimates in arid
conditions. Yet the microrelief itself may be due to various influences
including the activity of rodents and of certain types of vegetation.

The morphological structure of a landscape is not permanent or
invariable; it should be investigated from the genetic viewpoint. As it
develops, the structure of a landscape grows in complexity. Areas
occupied at first by a single facia (e.g. following marine regression)
differentiate with time into systems of facies and urochishcha. Nor
are the latter in any sense static. Most facies and urochishcha are
dynamic, which makes it difficult for us to categorically assign a
particular elementary geocomplex (an individual entity) to a particular taxonomic class. Thus a hardly noticeable rill develops into a
gully and ravine, and a small karst cone into a large karst basin.

The dynamic nature of the morphological units of landscape is not
only due to developing topography; the development of facies into
urochishcha, within a single relief form, may be due to vegetation,
leaching processes, etc. A classical example is provided by steppe
sinkholes (minor depressions) which convert during their evolution
from the elementary geocomplex (facia) into urochishcha with con-
centric facies comprising aspen stands, willow woods, and sometimes
oak stands as well.

N. A. Solntsev proposed the term 'geographic links' for such
elementary complexes which give rise to more intricate territorial
units.

The significance of the spatial relationships between morphological
units should not be exaggerated. The principal task of landscape
morphology is not to explain the mutual distribution of facies, urochishcha and sites, but to study their *interaction*, i.e. their effect on
the dynamics of heat, moisture, mineral and organic substances. The
morphological investigation of landscape should be regarded as a step
along the way to a better understanding of the processes occurring
within it. Such a study is based on the analysis of heat and water
budget and of the migration of chemical elements and it must use
geophysical and geochemical methods, but a necessary prerequisite is
the isolation, systematization and mapping of the constituent units of
a landscape, i.e. facies and groups of facies.

The facia as an elementary unit of a landscape

It has been demonstrated that the facia is the smallest indivisible
physical-geographic unit, and is characterized by uniform physical-
geographic conditions. Facies have been differentiated and mapped
for a long time by geographers, under different designations. The term

'facia' has numerous synonyms: the 'epimorph' (R. I. Abolin), 'elementary landscape' (B. B. Polynov and I. M. Krasheninnikov), 'microlandscape' (I. V. Larin), 'biogeocenosis' (V. N. Sukachev), etc. Foreign literature also includes a vast range of designations which essentially correspond to facies: the 'landscape cell' (Paffen), 'platform' (Schmidthusen), 'ecosystem' (Tensley), etc.

Contemporary Soviet geography uses the term 'facia'. It has been said that since this term is borrowed from geology, where it has a completely different meaning, it would be better not to use it in landscape science. It is true, of course, that the notion of facia originated in geology, nevertheless it does have a geographic sense. In geology facia designates a set of physical-geographic conditions producing sedimentary rocks; as shown by D. V. Nalivkin in 1933, this is quite consistent with the concept of the elementary geocomplex. 'The facia', wrote Nalivkin, 'is not only the sedimentary rock, i.e. a lithological complex, but a well-defined uniform part of a landform or of an oceanic bottom, i.e. a geographic or paleo-geographic concept.' Furthermore, 'facies include the biocenoses, the biotype and the sedimentary rock'. Finally, 'facia is a unit of landscape. All landscapes and the entire earth's surface divides into facies.'[17] These views were shared by N. M. Strakhov, L. V. Rukhin, G. F. Krasheninnikov, and other prominent geologists.

The view that there could exist geological facies which do not conform with geographic facies is erroneous. The supporters of this idea maintain that geological facies extend over larger areas than geographic facies. In fact, facies may vary greatly in size; some, as shown by D. V. Nalivkin, spread over large areas (e.g. the facia of the red deep-water clay in the Pacific Ocean) while others occupy tiny areas of a few square metres (e.g. oyster-banks). Naturally land facies are, as a rule, smaller than the underwater (especially the deep-water) facies.

L. S. Berg considered as unacceptable such ideas as geological facia, climatic facia, zoological facia, etc., since the term 'facia' implies the idea of a complex. Facies may be studied from different points of view, e.g. geological, geochemical, biocenotic, etc. According to Berg, the facia is 'an indivisible unit of geography, biogeography and geology.'[18]*

The major factor in the differentiation of facies in a given land-

---

* Thus the facia represents 'a geographic individual' to Soviet landscape scientists. In keeping with their traditional denial of the existence of concrete regions, Grigg (1967, p. 484) concludes that Western geographers generally do not recognize a 'natural' individual in the landscape. Rather, most Western geographers use 'operationally defined taxonomic units' (OTUs) in constructing a regional hierarchy 'from below'.

scape, i.e. in uniform zonal and azonal, climatic, geological-geo-morphological, biogeographic and other conditions, is the diversity of *localities*. In addition, the formation of various facies may be associated with the specific character of landscape development, such as the overgrowing of lakes, swamping of forests, etc.

We designate as a locality an element of relief (e.g. a section of a hill slope or valley, a peak, a foot of a hill, etc.) which is characterized by a specific relative elevation over the local base level, or by the exposure, the angle and the shape of the slope. In flat, upland areas the character of localities depends on the forms of microrelief and on their proximity to the lines of natural drainage.

The existence of small relief forms substantially affects the movement of the atmospheric masses; it changes the direction and velocity of winds near the ground, thus contributing to the redistribution of rainfall. The windward hill slopes usually receive less atmospheric moisture than the lee slopes. (Fig. 24) Especially important is the redistribution of snow; from the hilltops, and often from the windward slopes as well, snow is blown into depressions where it accumulates and remains one to two weeks longer than on elevated sections. Slopes create local air circulation; during the night, cool air descends into depressions, creating temperature inversion. According to I. A. Golts-berg, the frost-free period lasts twenty days longer on hilltops and upper sections of slopes than on flat open sections, and twenty-five days longer than in troughs.

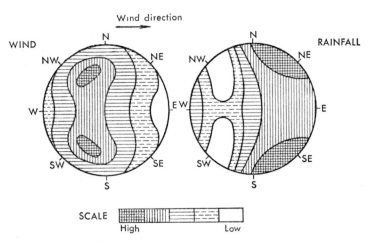

Fig. 24   Distribution of wind velocities and rainfall around a hill (after R. Geiger)

The exposure of the slope determines the distribution of solar energy (essentially of direct energy). In temperate latitudes, northern slopes receive less solar radiation annually than the horizontal surface, while southern slopes receive more. The difference is greater in the winter than in summer.

The gradient of the slope also affects the amount of solar radiation. During the winter when the sun is high above the horizon, steep southern slopes receive many times more energy than the flat slopes, while the steep northern slopes receive practically none. During the summer, on the other hand, the amount of radiation of steep southern slopes is reduced, since the sun's rays at midday fall on the earth's surface at an obtuse angle. At 50° N. latitude, southern slopes of 45° receive half the direct solar radiation of a horizontal surface.

Slopes not only affect the redistribution of moisture but also that of the solid matter over localities. The intensity of the runoff, and of the transport of solutes and clastic material, depends on the steepness and the shape of the slope. Depending on its exposure, the intensity of evaporation varies and this also contributes to the water supply conditions in each locality. If other conditions remain constant, the relative elevation of a locality determines the depth of groundwater.

Owing to the removal of soil particles and loose products of weathering in the upper part of the slope and their redeposition on flatter areas and at the foot, the bedrock undergoes certain changes, depending on locality. The mechanical composition and thickness of loose deposits, as well as the depth of the rock base, undergo variations.

Thus different localities are distinguished by individual heat and water conditions, and also with respect to the intake and outflow of mineral substances. This means that within a single landscape and over a uniform substrata, uniform ecological conditions or habitat conditions (ecotope) must characterize each locality (i.e. in another landscape different habitat conditions will occur in analogous localities, since the background of climate, geology and historical development, etc. will be different). In such conditions a *single* biocenosis will be established over an area corresponding to a single locality. Having settled in a particular area, organisms interact with their environment and actively transform it, producing soils and altering the original microclimate, water budget, and geochemical processes. It follows that with respect to biocenosis a habitat can be regarded conditionally, for analytic purposes, as a primary concept. The plant and animal population, together with the components of the inorganic environment attached to a given concrete locality, form the simplest geocomplex—the facia.

The facia constitutes the primary energy and geochemical cell in a landscape, similar to a cell in a living body. It is a unit from which we begin the study of the cycles and the conversion of energy and matter in a landscape. It must not be regarded, however, as an autonomous system with a closed cycle of matter and energy, since facies depend very closely on one another. Facies change regularly with the profile of relief, producing series of facies. A complete series of facies, including the positive and negative shapes of relief, connects two or more associated urochishcha. The study of a typical series of facies is fundamental to the understanding of landscape dynamics. The analysis of such series underlies the classification of facies.

B. B. Polynov established three principal types of 'elementary landscapes' or facies: *eluvial, superaqual* and *subaqual* (Fig. 25).[19]

Eluvial elementary landscapes are found on upland localities where the groundwater lies deep enough so as not to affect in any way the processes of soil formation and the vegetation cover. Materials enter only from the atmosphere, with rain and dust, but are lost both through runoff and through the downward flow of moisture into the ground. It follows that the outflow of material must exceed its intake. This situation is associated with the leaching of upper soil horizons and the development at some depth of an illuvial horizon. Because of the continuous removal of soil particles from the surface, the soil-forming process penetrates deeper and deeper into the underlying rock, gradually extending into new areas. Over a prolonged time-span, in the geological sense, a thick weathering mantle is produced which contains the residual (least transportable) chemical elements.

The vegetation on eluvial facies must struggle competitively against the continuous outflow of mineral elements. 'This struggle of two opposing processes—the intake of mineral elements by plants and their outflow from the soil by descending solutions—is characteristic of this category of elementary landscape.' Polynov stresses that 'the capacity of plants to assimilate mineral elements is responsible for the

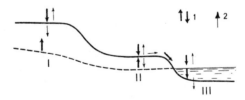

Fig. 25   Principal types of elementary landscapes (after B. B. Polynov)
Elementary landscapes: *I*, eluvial; *II*, superaqual; *III*, subaqual. *1*, entry of materials into landscape; *2*, elimination of materials from landscape. Dotted line shows the level of groundwater.

fact that even watershed soils in exceptionally wet regions are not completely leached with respect to any element'.[20]

Superaqual (above water) elementary landscapes are formed, according to Polynov, in localities where the groundwater flows close to the surface. This water rises to the surface as a result of evaporation and, together with various dissolved compounds, penetrates to the surface. As a consequence, the upper soil horizons are enriched with chemical elements which are characterized by a maximum migrating capacity (e.g. the solonchaks). In addition, substances may enter this horizon from the more elevated eluvial localities through transport by runoff.

Finally, subaqual (underwater) elementary landscapes develop along the bottoms of lakes and swamps. The material is supplied mainly by the runoff. The 'soil' (i.e. the bottom mud) accrues upwards and may be entirely unconnected with the underlying bedrock. Elements which are most mobile in these conditions accumulate in these 'soils'. Organisms inhabit a specific medium and include specialized biological forms. Underwater localities differ sharply from those on land with respect to the mineralization of organic residues and, as a result, sapropels* are developed in place of the humus.

It goes without saying that each of the three types of elementary landscapes enumerated is represented in nature by a multitude of variants in different zones, provinces and landscapes and within the latter by different urochishcha and facies. Moreover, there exists a vast range of intermediate types which also differ with respect to relief, substrata, etc.

It follows that B. B. Polynov's schema should be regarded only as a point of departure for a more detailed analysis; in any case it needs to be related to some concrete structures. Its importance lies in directing attention not so much to the purely external features of landscape morphology as to the interrelationships between the components of a landscape. The three types of 'elementary landscapes' proposed by Polynov constitute the principal links in a genetically associated series and exist in virtually every landscape.

A detailed classification of facies and localities must be developed for each individual physical-geographic province or group of similar landscapes. At the present time there are few examples of such classifications, and these relate only to tiny sections of detailed landscape surveys. K. G. Raman developed the system described in figure 26 of principal localities and facies applicable to the geography of the Latvian S.S.R.

* Humic mud, a deposit in lakes and lagoons; rich in plankton; grey-brown ooze.

Next, let us examine a typical facial profile of a southern-tayga landscape in north-western Ladoga. (Fig. 27) The natural (original) facies are subject to substantial changes resulting from human activity. For this reason the original facies of a particular type are represented

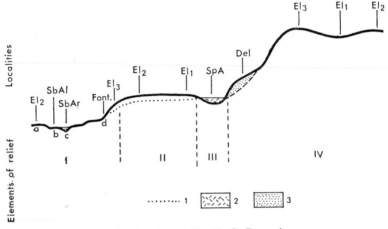

Fig. 26   System of major localities (after K. G. Raman)
I, valley; *II*, plain; *III*, basin; *IV*, hill. *a*, first terrace; *b*, floodplain; *c*, riverbed; *d*, valley slope. *1*, level of groundwater; *2*, peat; *3*, deluvial deposits.
Eluvial localities: *El₃*, steep slopes, intensive surface runoff, well-developed denudation of soils, a relatively xerophytic vegetation [adapted to dry conditions: ed.] (on sandy, outwashed plains and marine accumulative plains); *Cladonia* and heather-type fine forest (on carbonate-less morainic material) and *Hylocomium* spruce stands. El₂ gentle slopes and plains, with good drainage; typical upland localities, with cowberry pine forest (on sands) and *Hylocomium* and mountain-soil spruce stands (on carbonate-less moraine). *El₁*, poorly drained but not boggy plains, wet-forest types (*Molinia*-grass pine forest and *Molinia*-bilberry spruce forest) on gley and peat-gley soils. *El₁*, small watershed depressions either with traces of surface swamps (no contribution from groundwater) or with wet types with forest cover. *Foot of the slope localities*: *Del*, Deluvial localities on deluvium deposits, with pronounced eluvial characteristics (e.g. strong podsolization or carbonate deposition); often the area is improved by the transport of mineral and organic substances from more elevated parts; the vegetation comprises cowberry pine forests (on loose sands) and mountain-sorrel spruce stands (on carbonate-less moraine). *Font*, Fontinal (key) localities in areas where the mineralized groundwater breaks through to the surface, or in the vicinity of such places, and where the trophic conditions for organic life considerably improve the mineral nutrient supply; the vegetation consists of sedge-grass and reed-grass pine forest, and sedge-reed or fern-sedge spruce forest. *Depression and river valley*: *SpA*, Superaqual (above water-level) localities in areas with a shallow level of groundwater, marked swamp development, and covered with wild rosemary or peat-moss pine forest. *SbA*, subaqual localities (underwater seasonally or intermittently) (*SbAr*) in river courses (*SbAl*) on flood plains.

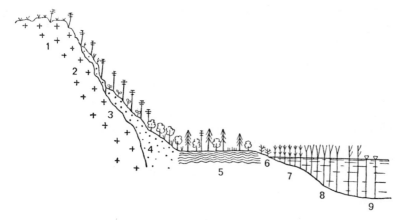

Fig. 27   Typical facial profile of a landscape in north-western Ladoga

1, rocky facies along the crests of selga ridges: fine crystalline Archaean granites or granite-gneiss outcrop; they create a complex microrelief, with rocky outcrops, micro-depressions, fissures and large detached masses; melkozem (fine earth) accumulates only in the depressions; moisture supplies are highly unstable. A mosaic moss-lichen vegetation retains the atmospheric moisture very poorly and grows on primitive accumulative soil only a few centimetres thick. Sparse dwarf-pine trees develop root systems along the rock fissures. *2, facies of steep (more than 10° to 15°) selga slopes*: a characteristic thin cover of coarse boulders and sandy loam, with intensive infiltrating runoff and uneven water supply. The vegetation consists of sparse pine forests, with cowberry, hair-grass and areas of bilberries. The soils are peaty-illuvial-humus, somewhat resembling mountain-tayga soils. *3, facies of lower, gentler (3° to 10°) selga slopes*: a thicker stratum of boulders and sandy loam, with light-admitting pine forest, together with bilberry, cowberry and hair-grass undergrowth as well as oak associations in the grass cover. Bilberry-spruce forest, with oak associations, often grows on northern slopes. The soils are humus rich, weakly podsolized and ferruginous. *4, facies of the selga foot*: a characteristic sandy-loam deluvium, abundantly supplied by runoff waters from more elevated areas, with speckled alder stands. The soils are soddy-crypto-podsolic, humus rich and ferruginous. *5, facies of inter-selga depressions*: mostly flat, sometimes closed (deprived of the natural surface runoff) and made up of heavy varve loams and clays. Depending on drainage conditions, facies with bilberry, mountain-sorrel or grassy speckled alder-pine-spruce forests (on gleyey soddy-medium-podsolic and humus-gley soils), together with sparse peat-moss and haircap-moss birch forest (on peaty-gley soils) are found, also moss-grass swamps. Many depressions are dry and overgrown with variants of mottley-grass vegetation, forming meadows and fallows. *6, shore facia*: carries hydrophytic vegetation and sparse willow stands along the normal level of the lake water-level. *7, shallow-water facia of overgrowing lake*: depth up to 0.5 m; closed *Equisetum heleocharis* vegetation. *8, shallow-water facia of overgrowing lake*: depth 0.5 to 1.2 m; closed *Equisetum heleocharis* and *Carex inflata* vegetation. *9, facies of open part of lake*: depth 2 to 4 m; reed-grass, white water-lily, yellow pond-lily and pond-weed vegetation.

at any given moment by a series of *derived facies* or, in L. G. Ramyenski's terminology, *modifications*, which correspond to the different
forms of human interference (e.g. burnt-off areas, tree-felling zones,
dry valley meadows, arable land extending over forest facies, etc.), or
to the various stages of return to the original facia following the
termination of human activities (e.g. fallow lands, waste-lands, areas
of scrub, areas with seasonal small-leaved forests, etc.).

Of course farming alters the plant cover and the animal population
of facies, as well as soils, water conditions and the microclimate. Thus
in place of the swampy forest facies in the inter-selga hollows
described earlier, we find, as a result of artificial drainage, ploughing,
introduction of fertilizers and pasture-grass cultivation, productive
facies with a much lower level of groundwater, cultivated soils, and
a completely different set of habitat conditions.

The modification of original facies has a more or less temporary
existence, however, and when human interference ceases there is a
return to a state close to the original. An exception is human activity
which alters the entire locality in such a way that the original facies
are completely destroyed; this is the case with building construction,
open-pit mines, water-storages, etc.

Productive modifications always exhibit differences due to the
type of base. Ploughing, for example, cannot completely eliminate all
the differences which arise, for example, from the variegated forest
cover prior to cultivation, or from differences in the exposure and
steepness of slopes. Classification should therefore reflect the connection between modification and the original facies.* This means that
the taxonomy of facies is always based on natural factors and the
associated basic facies, while modifications comprise, as it were, a
parallel series subordinate to the basic series of facies. For example,
an individual series may consist of pastoral modifications due to
intensive grazing, arable land, fallow and waste-land modifications,
etc.

Seasonal productive facies often result also from elemental natural
causes (e.g. fires, floods, rodent infestations, etc.). Finally, a separate

---

* This is undoubtedly one of the most serious problems in undertaking
physical-geographic classification. While an outline of attributes and variables
covering the 'natural' features of a place are easy enough to list and consider
when in the field, it is often quite difficult, if not impossible, to know exactly
which features have remained unaltered by cultural activities. In densely populated areas of the world soil and vegetation conditions have very often changed
to such an extent that they bear little resemblance to those of past centuries. It
is hypothesized that burning carried out centuries ago by Aborigines in parts of
Australia resulted in significant changes in vegetation characteristics. In situations such as these the reconstruction of natural conditions and even the understanding of what constituted 'the original facies' are most difficult tasks.

category in facial classification includes the so-called *serial* facies which constitute sequential series or stages in facia development under conditions of a relatively rapid course of geographic processes, e.g. on a mobile substrata, very steep slopes, on rocks subject to karsting, etc. V. B. Sochaya has suggested that we should, in such cases, classify not the individual facies, which are relatively short-lived, but their *genetic series*, e.g. the series of facies which develop on weathering outcrops of Lower Paleozoic limestones, or the series of alternating facies along the slopes subject to solifluction, etc.

The urochishcha, their geographic characteristics and principles of classification

The urochishcha are the most clearly differentiated parts of a landscape; characteristic groups of urochishcha provide us with the principal criterion for the establishment of landscapes in the field or from aerial photographs. They are especially well defined in dissected relief, with alternating convex ('positive') and concave ('negative') forms of mesorelief, where they correspond to the individual hills, basins, ridges, hollows, etc. The character of each urochishche depends not only on the shape of the landforms but also on their genesis and the lithological composition of the bedrock. Differences in substrata are reflected in the properties of the heat and water cycle, soil formation and biocenoses.

Differences in the character of the substrata along flat interfluves with uniform relief become the major factors in differentiating the urochishcha. Both the hydro-physical and the trophic properties of surface deposits are important in this connection, as well as the character of the underlying bedrock (e.g. water permeability, carbonate level, etc.). Along the flat or undulating plains in the area of Valdai glaciation a relatively variegated sequence of morainic (usually washed-out) and texturally heterogeneous (from sands to clays) glacial melt-water and lacustrine deposits are often found. This sequence of bedrock is virtually unmanifested in relief and yet it produces different urochishcha.

Different urochishcha may even arise when surface deposits are entirely homogeneous, i.e. in cases where these deposits vary in thickness and where below them at various depths are the Quaternary deposits or the bedrock of non-uniform lithological composition which directly affects the water cycle, soil formation and other processes. Thus typical forest urochishcha, with dry (lichen or cowberry) pine forest on ferruginous podsols, predominate over the thick lacustrine-glacial sands in the tayga landscape of the Russian plain.

Where, however, the sand horizon is thin and lies over impermeable moraine or varved clay, swampy urochishcha develop with haircap-moss and peat-moss forests and peat-moss bogs.

In differentiating urochishcha along the broad lowland interfluves another important factor must be considered; this is the natural drain-age condition, which depends on the distance between river valleys. As this distance becomes greater, swamp conditions occur due to the rise in the level of the watertable impeding runoff. This is reflected in the character of the soil and vegetation cover, and urochishcha alter in a direction towards the centre of flat watersheds. An example is provided by the middle-tayga landscape of the western Siberian plain which shows a typical series of urochishcha. (Fig. 28)

The processes associated with the development of a plant cover frequently also lead to the differentiation of a climatically and geo-morphologically homogeneous territory into urochishcha of different types. In cases where the environment supports competing plant communities equally well as happens in transitional zonal conditions

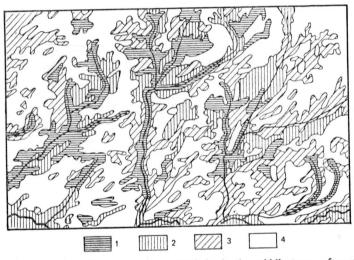

Fig. 28 Schematic distribution of urochishcha in the middle tayga of western Siberia (after V. B. Sochava, T. I. Isachenko and A. N. Lukicheva)
1, floodplain urochishcha, with dark-conifer, III–II site class forests, together with scrub undergrowth; *2*, riverbank urochishcha, with dark-conifer (mainly cedar-spruce) *Hylocomium*, IV class forests, together with productive birch stands; *3*, urochishcha of poorly drained interfluves, with peat-moss pine forest; *4*, urochishcha of the central parts of the interfluves, with surface peat-moss swamps.

(e.g. in the forest-steppe), a complex mosaic of communities of either type are found over a single locality. G. N. Vysotski drew attention to this phenomenon. Vysotski noticed that competing plant communities growing in a single area cover different sections of terrain, locally affecting the climate, water conditions and soils. As a result, urochishcha of different types (e.g. watershed forests and sections of grassland steppes) alternate without any apparent regularity.

Every urochishche represents a regular system of facies. The character of their association may differ considerably. Along the slopes the facies replace one another in sequence, forming elongated horizontal strips or rings. In these cases the urochishche embraces a part of the associated series of facies, consisting either of the eluvial facies (on convex landforms) or the superaqual facies (in concave relief). If differences in the physical-geographic conditions arise within a single urochische at different slope gradients, it will include not a single but two or more series of facies, each of which will constitute an individual sub-urochishche. Watershed urochishcha, the surfaces of which do not noticeably slope, are characterized as a rule by mosaic facial structure, with greatly reduced contrasts between individual facies.

Depending on their internal complexity, the urochishcha are divided into *simple* and *complex*.* N. A. Solntsev considers that in a simple urochishche each element of mesorelief corresponds to a single facia, while in a complex one each element corresponds to a system of facies, i.e. a sub-urochishche. For example, if only a single facia corresponds to each slope and the floor of a small valley, such a

---

* Christian and Stewart (1968, p. 248) place each of their land systems into one of three categories: simple, complex, or compound.
'In the early stages of its development it was found desirable to introduce three terms: simple land system, complex land system and compound land system. A simple land system is one composed of clearly definable land units which recur in association to form a simple recurring pattern such as an old peneplain with remnants of old drainage depressions, restricted to a climatic range which does not involve major vegetation or soil transitions due to the climate factor. A complex land system would be the combination of two or more such systems, for example, a similar peneplain uplifted and dissected at intervals by parallel streams, with two distinctive patterns, the peneplain pattern and the pattern of scree and colluvial dissection slopes, lower slopes, floodplains, stream levees and stream channels. The two simple land systems are geomorphogenetically related and this is the main distinction between the complex and compound land systems. The latter would be represented by a number of isolated igneous intrusions into a sedimentary area. The intrusions may represent outliers of a more extensive igneous land system elsewhere but are isolated within the sedimentary land systems with which they are less closely related geomorphogenetically than the components of a complex land system.' (Christian and Stewart, 1953)
See also Brink *et al.* (1966, p. 11).

valley is classified as a simple urochishche. If, on the other hand, several facies extend over each slope and the floor of a deep, large valley, we are in fact dealing with a complex urochishche.

One can, however, treat the idea of a 'complex urochishche' more broadly in view of the great diversity of structural forms manifested by this morphological unit of landscape. A few typical examples of complex urochishcha follow:

1. Large-scale mesorelief with 'entrenched' second-order meso-forms (e.g. a ridge with hollows or ravines, or a swampy basin with a lake).

2. A single, but lithologically varied form of mesorelief (N. A. Solntsev and his co-workers[21] propose an example where a single valley includes three individual urochishcha: (a) the crest—a semi-soddy, dry valley with loamy topsoil and with a morainic substrata; (b) the central part—a damp valley with creeping slopes exposing Jurassic clays; (c) the bottom part—a dry valley exposing Carboniferous limestones, with step-like slope structure).

3. A broad (dominating) watershed urochishche with small fragments of second-order urochishcha or single facies (e.g. swamps, basins, karst, zoogenic, etc.).

4. 'Double', 'treble', etc. urochishcha (e.g. a system of merging swamp regions, each of which constitutes an individual simple urochishche).

With respect to their place in the morphological structure of landscapes, the urochishcha may be divided into *primary* or *dominant* and *secondary* or *subordinate*. The former constitute a type of base landscape, usually occupying the major part of the area, while the latter play a secondary role. It should be noted, however, that as yet no precise criteria for differentiating these categories have been determined. The staff of the Landscape Laboratory of the Moscow State University distinguish among the basic urochishcha a dominant group which occupies most of the area and creates a landscape 'background', and a subordinate group spread against the background; the latter urochishcha have a fairly high incidence but occupy smaller areas, i.e. they usually characterize suffusion depressions, valleys and other negative relief forms. The secondary urochishcha differ from the primary but only by their lower incidence. (Fig. 29)

The classification of urochishcha with respect to their area is important for the description of concrete landscapes, but it cannot form the basis of the classification of urochishcha since the areal relationships between them vary; the same group of urochishcha may be dominant in one landscape, and subordinate in another. It follows from this that such a classification is both relative and somewhat conditional.

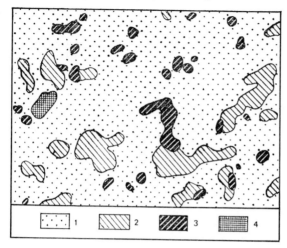

Fig. 29 Schematic relationships between the primary and secondary urochishcha on the Oka bank terrace-plain landscape (after N. A. Solntsev et al.)
Background, dominant urochishche: *1*, flat terrace above floodplain level, consisting of 3 m layer of sands over loams, with soddy-podsolic sandy soils and
green-moss pine forest or secondary birch forest (partly under cultivation).
Subordinate urochishcha: *2*, damp depressions, with sands overlying loams at a
depth of less than 2 m, soddy-podsolic gley and gley sandy-loam soils, with
spruce-pine and aspen-pine haircap-moss bilberry forests; *3*, scrub cotton-grass
sphagnum and sedge-sphagnum swamps with transitional peat bogs. *Secondary
urochishcha*: *4*, terrace subsidences, consisting of loams with soddy-podsolic
gleyey, and gleyey weakly to moderately podsolized soils, with spruce forests
and oak forests mixed with pine.

In classifying urochishcha one must proceed from their *genetic
similarity** or *differences*, and must take into account the associations
of facies characteristic of every type or species of urochishche. From
this point of view it is necessary to distinguish two major categories:
1. Urochishcha associated with convex mesorelief and with upland
catchment areas, characterized by good drainage, deep watertable,
downward movement of moisture and solid material, and a predominance of eluvial or automorphic† facies.

* Genetic similarity is considered when land facets are grouped into land
systems. (Brink *et al.*, 1966, pp. 10-11) One definition states that a land system
'is a recurrent pattern of genetically linked land facets . . . In a compound land
system the constituent facets fall into two or more groups. Within each group
the facets are cognate (genesis consistent with that of the land system); they are
genetically linked'.
† See discussion concerning B. B. Polynov on pages 142-3.

2. The urochishcha of concave mesorelief (e.g. erosion, subsidence, karst and other origins); also low-lying terraces the water supply of which is supplemented by surface runoff, often accompanied by the capillary rise of groundwater (in arid regions this rise is associated with the accumulation of salts). These urochishcha have characteristic local-climatic features (e.g. downward movement of cold air) and consist mainly of superaqual or hydromorphic,* as well as subaqual facies.

A further classification takes account of the genetic forms of relief, the lithology and mechanical composition of bedrock, and the character of water supplies and drainage. Even more detailed classification of drainage basin urochishcha is based on lithology, together with the genesis of relief forms and the degree of dissection. The classification of the second category of urochishcha must be based on such factors as the presence or absence of runoff, flowing or stagnant water supplies and the type of water source (e.g. atmospheric, ground, flood, etc.).

Different landscape zones, subzones and provinces on uniform relief and bedrock give rise to different local climates, moisture, soil and biocenotic conditions, i.e. to different urochishcha. For example, drainage basin urochishcha on morainic plains in the southern tayga are characterized by the predominance of *Hylocomium* spruce forests (bilberry and mountain-sorrel) and podsolized soils; in the subtayga (the zone of mixed forests) the same urochishcha show a predominance of complex (under-storey) spruce forest or broad-leaved spruce forests on soddy-podsolized soils. For this reason the classification of urochishcha, as well as of facies, must take account of their zonal and provincial characteristics. Each type of urochishche, as a rule, must be isolated from a single landscape province.

An example is provided by the following schematic classification of urochishcha of the southern tayga in the north-western part of the Russian plain.

Hill and ridge, intensively drained urochishcha

1. *Morainic hills and ridges* (mainly carbonate-less boulder loam), with spruce forests of different types (mainly bilberry and mountain-sorrel) on weakly to moderately podsolized soils.

2. *Kame hills* composed of sands and sandy loam, with lichen and *Hylocomium* pine forest (mainly cowberry) on surface-podsolized soils and ferruginous podsols.

---

* See discussion concerning B. B. Polynov on pages 142-3.

Urochishcha of elevated and well drained interfluves

3. *Interfluvial plateau on limestones and products of limestone weathering* on thin carbonate moraine, with complex spruce and broad-leaved spruce forests on soddy-carbonate, leached and podsolized soils.

4. *Interfluvial plateau and elevated plains* consisting of thick moraine with underlying carbonate bedrock, with complex spruce forests on soddy-carbonate, podsolized, and soddy, weakly-podsolized soils.

5. *Spurs and undulating interfluves of elevated morainic plains* consisting of medium or light carbonate-less boulder loam, with *Hylocomium* spruce forest (bilberry and mountain-sorrel) on moderately or weakly podsolized soils.

Urochishcha of low-lying undulating, moderately drained
(dry) interfluves and terraces

6. *Low-lying interfluves* on carbonate moraine or shallow-lying limestones, with floristically rich mountain-sorrel spruce forests or complex spruce forests on soddy, weakly podsolized or soddy-carbonate podsolized soils.

7. *Low-lying interfluves* on carbonate-less washed-out moraine, often mixed with sand or covered with thin sandy loam, with *Hylocomium* spruce forest on strongly podsolized and, in places, gleyey soils.

8. *Low-lying terraced interfluves* on lacustrine, carbonate-less clays and loams (and, in places, alluvium), with *Hylocomium* spruce forest on strongly or moderately podsolized and, in places, gleyey soils.

9. *Terraced interfluves* on lacustrine sands and loams, with pine forest (mainly cowberry) on surface-podsolized, ferruginous soils.

10. *Frontal, apron interfluves*, with lichen and *Hylocomium* pine forest on sandy, surface-podsolized, ferruginous soils.

Urochishcha of low-lying, flat, swampy terraces and
interfluves

With hard groundwater:

11. *Low-lying interfluves* on rewashed boulder loam and partly lacustrine loam, with swampy grass and scrub spruce forests on soddy-gley soils.

With surface or moderate groundwater supplies:

12. *Low-lying interfluves* on washed-out, sandy boulder loam, with swampy (mainly haircap moss) spruce forests on podsolized-gley and peaty-podsolized-gley soils.

13. *Low-lying terraced interfluves* on lacustrine clays and loams, with swampy spruce forest on podsolized-gley and peaty-podsolized-gley soils.

14. *Low-lying terraced interfluves* on lacustrine sands and sandy loams, usually on shallow-lying moraine, with haircap-moss and peat-moss pine forests on peaty podsols.

Swampy urochishcha on flat, undrained interfluves and depressions

15. *Low-lying grassland and grassy swamps.*
16. *Intermediate peat-moss and sedge-grass swamps.*
17. *Peat-moss swamps.*

Urochishcha of poorly drained basins

18. *Lacustrine basins* on clays and loams, with *Hylocomium* spruce forests and tall grass on swampy, weakly podsolized, gleyey, humus-gley and peaty-podsolized-gley soils.

19. *Intermorainic basins*, with swampy spruce forests on peaty-gley and peaty-podsolized-gley soils.

20. *Interkame basins*, with wet and swampy pine forest on peaty-podsolized, ferruginous, humus and partly ferruginized soils.

Urochishcha of river valleys and floodplains

21. *Deep trench valleys*, with steep slopes and different types of forest (mainly spruce) on soils with varying degrees of podsolization.

22. *Small-river and stream valleys*, with tall grass and partly swampy spruce forest on humus-gley and humus-peat-gley soils.

23. *Floodplains* on clayey and loamy alluvium, with floodplain forests, meadows and scrub on soddy-alluvial meadow soils; also low-lying swamps.

Urochishcha of coastal beaches and dunes

24. *Sea coast beaches*, without vegetation or with psammophytes; also dunes, with grassy pine stands.

Lacustrine urochishcha

25. *Lakes of different types.*

The above classification has been developed for use in landscape mapping at a specific scale (1 : 600 000 to 1 : 1 000 000) and is far from exhaustive. It includes only the higher taxonomic units, i.e.

types and subtypes of urochishcha. Some unmapped urochishcha are not considered at all, while most of the other types could be differentiated further by taking account of relief meso-forms (e.g. large- and small-hill morainic urochishcha), dominating slopes, local climate and other features specific to each type, especially in the case of the valley, basin, swamp and lake urochishcha. The classification includes only the basic urochishche, i.e. taking no account of man's interference.

To illustrate the above classification, two cartographic examples are provided. The first (Fig. 30) represents a small section of the landscape of north-western Lake Ladoga, together with characteristic types of urochishcha; the other (Fig. 31) illustrates the types of urochishcha along the lacustrine-glacial sandy plain in the middle tayga.

The principal natural features of mountain landscapes

Mountain landscapes differ from lowland landscapes by their much greater complexity; this is due to their distinct vertical differentiation, which is superimposed over the horizontal. The most characteristically distinctive feature of mountain landscapes is their vertical stratification, discussed earlier in chapter 2. This stratification erases other

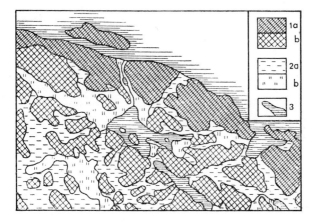

Fig. 30 Schematic distribution of the major types of urochishcha in the north-western Ladoga landscape
1, selga urochishcha on acidic crystalline rocks, with predominantly pine forest on weakly and partially podsolized soils; *1a*, high Ladoga-bank selga, with a cooler local climate and large rock facies; *1b*, low-level, intra-bedrock, and less rocky selgas. *2*, inter-selga urochishcha on banded loams, flat and poorly drained; *2a*, with green moss and tall grass, with spruce, pine, birch and speckled-alder stands, and relicts of transitional moors; *2b*, cultured, artificially drained, predominantly with mottley-grass meadows. *3*, lakes.

Landscape Science

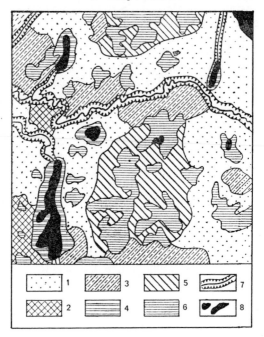

Fig. 31   Schematic distribution of urochishcha types on a section of mid-tayga lacustrine-glacial plain (after Z. V. Dashkevich)
1, gently hilly, sandy plain with spruce-pine, lichen and *Hylocomium* moss, bilberry-cowberry forests on well-drained, weakly and moderately podsolized soils; *2*, kame urochishcha with lichen-heather and moss-lichen bilberry-cowberry pine stands on well-drained top- and weakly-podsolic soils; *3*, small-hill plain urochishcha on twin-layer deposits, with gentle slopes, moss-cowberry and moss-bilberry stands on medium and strongly podsolic soils; *4*, lake-basin urochishcha with rain-type spruce forest (wet-grass, large-fern, haircap-moss and bilberry) on podsolized soils (gleyey and peaty in equal degrees); *5*, flat, sandy plain urochishcha with green haircap-moss moss-bilberry and haircap-moss sphagnum spruce-pine forest on swampy soils; *6*, highland and transitional moors; *7*, river valley urochishcha with poorly defined floodplains and elevated terraces, with spruce and spruce-pine forests of different types (from wet-grass and haircap-moss to dry-lichen); *8*, lake urochishcha.

geographical mechanisms in mountain regions so strongly that the latter are often forgotten when mountain landscapes are investigated. Until recently vertical stratification was usually considered in isolation from the more general (zonal and azonal) factors of geographic differentiation, which continue to act in mountain regions just as they do on the plains. And just as vertical stratification occasionally commands too much attention, other factors which determine the physical-geographic differences between individual mountain landscapes command too little.

It must be noted that mountain landscapes have not been nearly as exhaustively studied as lowland landscapes. For this reason the problems of structure, morphology, classification and regionalization of mountain landscapes have not been adequately elaborated. The idea of 'mountain landscape' itself lacks a clear definition. The relationship between landscape and vertical stratification is open to discussion. Are these concepts in fact valid, or should landscape perhaps be regarded as a section of the stratum? Alternatively, could a landscape extend beyond a single stratum, perhaps to embrace an entire series of strata?

Before we attempt the solution of these problems, the relationship between the horizontal and the vertical physical-geographic differentiation of mountains must be examined. Real differences not only in the vertical but also in the horizontal plane are observed. These differences spring from zonal and azonal factors, as well as from local factors which are primarily orographic in origin. A strong dissection of relief produces great variety and a contrastive distribution of mountain landscapes and their morphological components. These alter the real character of the vertical stratification, resulting in numerous local variants.

Among the orographic factors, notwithstanding absolute altitude, the major factor is slope exposure or aspect. Depending on the magnitude of the physical-geographic action, two types of slopes are distinguished: the principal slopes of a mountain range, oriented relative to its general direction and resulting from orogenic movements; and minor (secondary) slopes, produced mainly by the exogenous dissection of mountains and sometimes by specific features of their tectonic structure. Owing to the meriodional direction of the Urals, their principal slopes face east and west; the macro-exposures of the Great Caucasus face north-north-east and south-south-west.

The general direction of the strike of large mountain ranges and the presence of two main macro-exposures play an important role at higher levels of physical-geographic differentiation. The drainage divides of high ranges, e.g. the Great Caucasus, constitute important climatic boundaries; the boundaries of the latitudinal zones and the longitudinal section of continents often coincide with these boundaries and vertical stratifications of different types develop along the opposing slopes.

Secondary slopes are responsible for substantial contrasts in climate and microclimate, creating local variants of vertical stratification against the dominant background. The effects of these secondary factors underlie the differentiation of individual landscapes and, within these, that of their morphological units. The effect of slopes on climatic features makes it necessary to distinguish two types of exposure: solar or insolation and wind or circulation exposures. The former account for the orientation of the slope with respect to solar rays, and the latter with respect to atmospheric currents.

Solar exposure determines both the heat and the water conditions along the slope. Southern slopes receive more heat and are subject to more intensive evaporation than northern slopes and so, other conditions being equal, they should be generally drier. This situation is reflected in vertical stratification. Along southern slopes, boundaries of strata are usually displaced upwards, by comparison with those along northern slopes; in the steppe zone the forest stratum may be altogether absent along the southern slopes. In steppe Transbaykal, for example, the northern slopes are often under forest whereas the southern slopes (Solnopyeki) are covered with steppe vegetation. In temperate latitudes, where the sun remains relatively low over the horizon, the significance of solar exposure in the differentiation of mountain landscapes is the greatest. In the arctic and the tundra, slopes of all exposures are illuminated during the polar day, and because of the low position of the sun over the horizon, differences in the illumination of slopes are very small. At the same time, in low latitudes, where the sun rises high above the horizon, differences in the insolation of slopes of various exposures are likewise very small.

Wind exposure has a dual significance. On the one hand, it may lead to differences in the thermal conditions of opposing slopes; this is most characteristic of ranges with approximately latitudinal orientation (e.g. the Crimean mountains, the Great Caucasus, the central Asian ranges, etc.). The northern slopes of such ranges are subject to the action of cold air masses, whereas the southern slopes are more or less protected from them. As a result, exposure to cold or hot air currents intensifies the effect of solar exposure.

The other effect of wind exposure on the climate of slopes has to do with their orientation to the movement of wet air masses and cyclones. It has been noted in chapter 2 that the windward slopes of mountain ranges frequently receive much more rain than the lee slopes. In temperate latitudes the western and the northern slopes receive most rainfall. In subtropical latitudes the picture may be reversed. For example, the southern slopes of the Alai system of ranges, mainly the Gissar Range, receive increased rainfall from the cyclones of the Persian polar front during the spring when it moves from south to north. Accordingly, maximum rainfall in this area falls along the southern and south-western slopes, whereas the slopes on the opposite side of the range exhibit foehnic (rain-shadow) features.

The effect of wind exposure in relation to moisture-bearing air currents must be particularly strong in dry-climate conditions, although it is manifest in all zones, including the equatorial. In addition, the differences in rainfall on windward and lee slopes differ with altitude. Since increased rainfall is most noticeable along one slope the greatest contrast is found at levels corresponding to the most intensive rainfall area. In the Tyan-Shan this stratum lies approximately 1500 to 3000 m above sea-level, and in the south up to 4000 m. Along the northern and western slopes at these levels forests appear which do not exist on the opposite exposures. Further up and down, the differences in rainfall between the windward and lee slopes equalize to some extent. This is shown, in particular, by the difference in the snowline level which varies in the Tyan-Shan between 500 and 800 m. The existence of cold deserts in the eastern Pamir at the level of 4000 m, is due to its location in the barrier shadow of still higher ranges.

The effects of the wind and solar exposure often reinforce one another, and in such situations the contrasts in the climates and landscapes of opposing slopes become especially noticeable and are distinctly reflected in the structure of vertical stratification. The more humid and northern character of the strata along the northern and western slopes of the Altai or the Tyan-Shan, by comparison with the southern and eastern slopes, can be attributed primarily to the 're-capture' of the moisture transported by the westerly airstream. In addition, however, the inadequate insolation of the northern slopes, and their exposure in the lower parts to cold air intrusions (Fig. 32) contribute to the overall effect.

As they create or intensify the contrasts in physical-geographic conditions, exposure factors produce a situation in which the zonal and provincial features of vertical stratification largely become subordinate to the general plan of the orographic structure of the given mountain region. Thus the strata along the western and eastern slopes of the Urals (within specific landscape zones) show individual provincial features. The landscapes along the eastern slopes have a more continental, and in the southern part of the range a more arid, character; the strata of broad-leaved and dark-conifer forests found along the western slopes of the southern Urals are absent along the eastern slopes.

The Great Caucasus provides an even clearer example of the extent to which the distribution of the zonal type of stratification and its

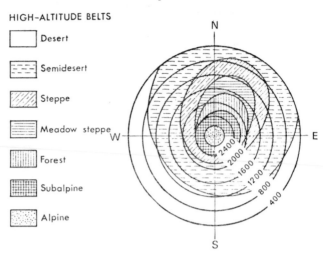

Fig. 32   Effect of exposure on the high-altitude stage on the Transili Ala-Tau Range (after O. E. Shchukina)

variants depends on orography. The drainage divide of the Great Caucasus separates the steppe-type belts along the north-north-eastern slope from the various subtropical types along the south-south-western slope. The low, lateral Suram Range in Transcaucasia sharply divides the wet subtropical mountain landscapes, with their characteristic stratificational spectra, from the arid landscapes of eastern Trans-caucasia.

The steepness and the shape of slopes constitute additional factors in physical-geographic differentiation. Increasing gradient is accom-panied, as a rule, by stronger contrasts in climatic conditions, owing to more intensive precipitation along the windward slopes and increased aridity along the lee slopes. Also important is the mutual orientation of the ranges: in complex mountain systems consisting of a series of parallel ranges and transverse spurs, the outer ranges receive more rainfall than the inner ranges, which lie beyond the influence of the moisture-carrying atmospheric masses, i.e. they lie in a kind of barrier shadow.

Intermontane depressions occupy a special position in the system of geocomplexes in mountainous regions. Closed depressions sur-rounded by ranges nearly always have a drier continental climate than mountain slopes at the same altitude, and they exhibit strata spectra characteristic of more arid zones. The most typical in this regard are large tectonic depressions in the mountains of the desert-steppe zone (e.g. the desert sinkholes in Tyan-Shan, the mountain-steppe basins in the Altai, etc.). Landscapes of longitudinal depressions lying in the barrier saddle between the Skalisty and the Bokovy Khrebet Ranges of the Great Caucasus have a relatively arid character. In this area widespread steppe and mountain-xerophytic plant communities and pine forests are found, whereas along the outer slopes beech forests predominate at the same altitude.

Some intermontane depressions (e.g. the southern Urals, Trans-baykal, etc.) exhibit inversions of vertical strata, i.e. a reverse sequence in the vertical plane. This phenomenon typifies small tectonic basins and broad mountain valleys in the temperate zone with a sufficiently wet climate but a frequent recursion of the anti-cyclonal atmospheric conditions, facilitating the descent of cold air along the slopes. The floor of such a depression in the tayga zone often exhibits tundra-type complexes which are replaced higher up with sparse forest, which in turn give way to mountain tayga.

And so the complex orographic dissection of mountain regions is responsible for a multitude of local variants of the vertical-stratifica-tional spectrum. Relief causes displacements in the vertical boundaries, some strata increase in size, others become smaller; some strata disappear altogether and the standard sequence may become inverted.

The next important factor in physical-geographic differentiation is geological structure, which is more complex in mountainous regions than along the plains. In mountains the substrata change more frequently with respect to lithological composition and position. Unlike the plains, mountains are composed essentially of dense rocks, sedimentary and igneous, which constitute the immediate substrata for soil formation and for the plant cover.

The effect of bedrock on other landscape components is, in general, well known. Thus the soil-forming processes are not identical on the sedimentary and massive-crystalline rocks; the latter undergo much more intensive changes in consequence of weathering and soil formation. Acidic and basic rocks, respectively, have a different effect on the migration of chemical elements, and on all associated pro-cesses. The podsolizing process on acidic, massive-crystalline rocks rich in silicon attains full development; the vegetation cover on these rocks also exhibits certain specific features (e.g. the 'granitic forests' in the Kazakh Fold highlands; the pine's penchant for granitic outcrops has also been observed in the Urals, and in other areas). On basic rocks, soils are richer in humus; an example is provided by the chernozem soils of the mountain steppes on the Armenian volcanic plateau, which are famous for their fertility.

Among dense sedimentary rocks, carbonate rocks are distinguished by their dramatic effects on the different components of landscapes. Even in typical mountain-tayga conditions they do not give rise to podsolization; the plant cover includes the characteristic calciphilic forms.

The relationships between typical forms of mountain relief and geological structures, rock density, solubility, jointing and other physical-chemical properties, are so well known that there is no need to consider them in any detail.

One example of the effect of geological structure on the formation of mountain landscapes is provided by the two stage structure of the main (southern) ridge of the Crimean mountains. The lower stage consists of clay shale and Taurian-series sandstones (Lower Jurassic to Upper Triassic) and Upper Jurassic coral limestones. The clay shales are easily abraded and form gentle slopes. The limestones create a level crest plateau (the yaila, a series of monoclinal limestone plateaux dissected by karst valleys) with steep discontinuities facing the Black Sea coast. Owing to jointing and the high solubility of the limestones, rainfall penetrates downwards, producing multiple karst forms, and the yaila lacks surface drainage. Its dryness accounts for the lack of forests. Landslides characterize the lower-stage shales. Owing to the impermeability of the shales, groundwater emerges at the surface along the interface with the limestones, giving rise to numerous streams. In normal conditions the Taurian-series slopes are covered with forest. Accordingly, the line dividing the Taurian-series outcrops and the Upper Jurassic limestones constitutes an important physical-geographic boundary; on either side of this boundary different landscapes prevail.

Relief and the geological structure of mountains change not only in a vertical but also in a horizontal direction and these changes are interconnected. The traditional classification of mountains as low, medium and high expresses not only the changes in relief with height but also directional changes from the centre towards the periphery. This classification defines three principal stages which reflect the history of the formation of a range, the age of its individual parts, the intensity of tectonic movements and the character of exogenous dissection.

The periphery of large mountain ranges consists predominantly of low foothills (e.g. the 'benches' along the northern slopes of Tyan-Shan), and low frontal ranges or spurs of major mountain chains; in arid climates these foothills are often strongly dissected by seasonal watercourses, with tracts of piedmont proluvial deposits. The next stage is the mid-mountain area which typically exhibits erosive dissection, and relatively gentle, rounded mountain shapes. Finally, the axial zone consists of high-altitude peaks characterized by glacial-nival topography, alpine structures and glaciers.

These steps or stages are often apparent in the geological structure of a range, i.e. in the sequence of strata of different ages, lithological composition, and degree of dislocation. Once again a good example is provided by the Great Caucasus. The lower, peripheral part of the range consists mainly of Neogenic, sandy-clayey deposits and con-glomerates, which constitute the low-mountain stage. Above it there is a sequence of older rocks the age of which increases progressively: Paleogenic, Cretaceous, Jurassic and Paleozoic. The highest, axial part of the range (in the western part) consists essentially of acidic, massive-crystalline rocks.

Superimposed over the vertical stratification, this step-like geologi-cal and topographic structure complicates the physical-geographic differentiation of mountains. The altitude strata do not necessarily coincide with stages in the above sense, yet they are obviously inter-related. The three principal stages outlined above are also reflected in climate. Climatic conditions over the foothills are closely associated with the atmospheric circulation over the plains. It is precisely in this zone of our southern mountains (up to 500 to 600 m altitude) that the effect of cold air intrusions from the north is manifested. In the mid-mountain stage the air masses rise and fronts develop; as a result, slopes facing the prevailing winds receive the maximum rainfall. Here there are particularly sharp climatic contrasts between slopes of opposite exposures. The high-mountain ranges (e.g. above 2000 to 2500 m in the Great Caucasus and above 2500 to 3000 m in central Asia) lie in the zone of ascending, free atmospheric currents; the landforms below have no significant effect on climate, being virtually independent of the ground layers of the atmosphere around and above the range.

Therefore a direct connection between the vertical stratification and the three stages described is found. The lower stage corresponds, as a rule, to the bottom (first) stratum which transforms directly into low-land landscapes. The mid-mountain stage exhibits the most complex and diverse vertical stratification; here at the same altitude are often found sequences of different strata. In desert-zone mountains this stage coincides with the forest stratum, although, this stratum is never continuous. Conditions are more uniform in the high-mountain stage. It is true that the lower part of this stage still reflects the latitudinal and longitudinal orientation of the range. For example, the tayga high-mountain stages are characterized by a mountain tundra stratum, while in more southerly zones the latter is replaced with mountain meadows. Above a certain level, however, which is not the same for all mountain landscapes, the differences in vertical stratification are virtually erased. Theoretically, the vertical stratification should terminate with a glacial-nival stratum, and it is only because many ranges do not reach the snowline altitude that this stratum is not always present.

It is concluded that the *stage sequence* of mountain landscapes must be regarded as an important complex index of the horizontal and vertical differentiation of mountains. The concept of stage sequence has a broader and more complex meaning than the concept of vertical stratification, which is based essentially only on soil and geobotanic criteria. The classification of mountains into low, medium and high is hardly ever done on narrow, geomorphological grounds; it obviously has an extended, complex physical-geographic meaning which still awaits precise definition. The concept of stage sequence for mountain landscapes includes such factors as their hypsometric position, principal geomorphological features and the character of their vertical stratification. The idea of landscape stage includes, as already noted, all the principal features of both vertical and horizontal differentiation; this is not the case with regard to vertical stratification alone.

By contrast with strata, landscape stages have a universal character. They are valid for the landscape classification of all mountain ranges. Every mountain system, provided it reaches the appropriate altitude, show a distinct lower, middle and upper stage. Vertical stratification, on the other hand, does not have a general character; many strata have only a local significance and spread over a limited area.

It follows that classification by stages enables us to compare mountain landscapes in different zones or regions, and facilitates the classification of mountain landscapes. Each stage embraces a part of the vertical spectrum of strata with a fairly uniform genetic history. A single stage may include several strata which replace one another not only in the vertical but also in the horizontal direction, depending on geographical position and local orographic factors, along slopes of different exposures, etc. The elevational strata belonging to the same stage in different landscape zones, regions and provinces are subject to a kind of geographic alternation (e.g. mountain tundra, mountain meadows and desert-steppe highlands).

Stage classification reflects the degree of development or the completeness of vertical stratification. In low mountains, i.e. those contained within a single stage, only the first stratum is generally found. High mountains, i.e. those which exhibit all three stages, exhibit a full sequence of strata. Medium-height mountains comprising two stages occupy an intermediate position.

Three-stage classification is the most typical system. It is possible
that it will be necessary to classify certain ranges in still greater detail.
For some areas, on the other hand, even three stages are excessive
(e.g. mountain ranges in the Arctic). In lower latitudes the complexity
of stage classification increases and so does the contrastive character
of stages. Moreover, the upper boundaries of each stage are not
everywhere firmly established; they depend on the geographic position
of the range, especially on the landscape zone, and on their history of
development. A more detailed development of the stage classification
of mountain landscapes is a matter for future research.

It is necessary to return to the concept of the mountain landscape
in order to establish its criteria. It would obviously be incorrect to
deny mountain landscapes conformity with an elevational stratum or,
alternatively, to regard them as a part of that stratum. A stratum does
not constitute a single entity, either in the genetic or territorial sense;
as a rule it represents tracts of land which differ in nature and occur in
isolation. For example, the dark-conifer mountain-tayga stratum in the
northern Urals embraces in a continuous belt the lower part of the
range; in the Altai this stratum is no longer continuous but occurs
only in the wettest peripheral part of the mountain, mainly in the mid-
mountain belt; in the Tyan-Shan the mountain-tayga stratum consists
of individual fragments wherever conditions are particularly favourable (e.g. along the northern slopes and in gorges). It is necessary to
add that the history of each of these mountain ranges differs considerably, so that the geological structure and the soils in the corresponding
mountain-tayga strata are also quite different; even the uniformity of
the plant cover is highly questionable.

On the other hand, it is known that fragments of very different
strata (e.g. forest, steppe or meadow) frequently occur side by side
along the same altitude but in different exposures, producing a
complex spatial mosaic. Strictly speaking, vertical stratification laws
should form a part of intra-landscape (morphological) differentiation,
since sections belonging to a single stratum (e.g. identical urochishcha
along the plains) recur in different mountain landscapes. We recall
that vertical stratification is not the only law underlying the physical-
geographic differentiation of mountain ranges, but is an active
phenomenon which is subject to the influence of latitudinal zonality
and longitudinal provinciality.

A more important index of the physical-geographic differentiation
of mountains is the structure of vertical stratification, i.e. the entire
series of strata comprising a concrete section. I. S. Shchukin and
O. E. Shchukina have found that landscape regions established with
reference to structural differences in vertical stratification in different

sections of the range simultaneously reflect horizontal and vertical differentiation.[22] Areas characterized by specific (individual) strata spectra, i.e. from the bottom up, however, can hardly be regarded as a single landscape, since a complete series of strata from the foot of the range to the peaks of its highest mountains is spread over different stages, diverse types of bedrock and changing climatic conditions. Explicit accounts of entire series of strata are of considerable importance in the physical-geographic classification of mountains, but only at higher levels of regionalization, i.e. right up to the level of districts.

It seems more correct, therefore, to differentiate landscapes within individual landscape stages. More precisely, a mountain landscape constitutes a part of the landscape stage within the boundaries of an independent (local) system of vertical stratification, and is characterized by structural-lithological and geomorphological uniformity.

Many mountain landscapes do not extend beyond the limits of a single elevational stratum, but they sometimes embrace associated sections of different strata or include, against the background of a single dominating stage, fragments of other stages as well. One example is provided by the landscapes of Soviet Transcarpathia. (Fig. 33) The landscape of the Uzhgorod-Zhust volcanic range is a low-mountain type, lying in a beech forest stratum. The landscape of the western Polonina region of the Carpathian mountains is a mid-mountain type. Here beech forests are also found but this stratum already represents a transition to a mountain-tayga stratum; fragments of the subalpine stratum are encountered. The Chernogorye landscape also belongs to the mid-mountain type but is more complex than the others; it consists mainly of mountain-tayga strata, the lower strip of which shows a transition to beech forest stratum, while the upper strip comprises a narrow subalpine stratum above which are even found fragments of alpine stratum. It may be that the upper stage of the Chernogorye, which includes a subalpine stratum and fragments of an alpine stratum, should be differentiated into an independent landscape.

The intermontane basins represent independent landscapes (e.g. the Vyerkhovin valley and the Marmarosh basin in the Carpathian mountains). Large tectonic basins, however, with heterogeneous climate, soil and plant cover, and other features, consist of different landscapes (e.g. the Issyk-Kul basin shown in figure 8).

The morphology of mountain landscapes is far more complex than the morphology of lowland landscapes. The system of morphological units used in differentiating lowland landscapes is clearly inadequate for mountain ranges, although the basic categories, facies and uro-

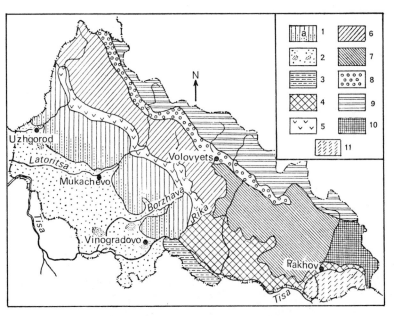

Fig. 33 Soviet Transcarpathian landscapes (after V. B. Sochava)
1, Uzhgorod-Ust Range landscape; *1a*, urochishcha of the foothills zone; *2*, Pritisen low-lying landscape; *3*, Transmarmarosh volcanic-ridge landscape; *4*, Marmarosh basin landscape; *5*, Predpolonina longitudinal valley landscape; *6*, western Polonina Carpathian landscape; *7*, eastern Polonia Carpathian landscape; *8*, Vyerkhovina longitudinal valley landscape; *9*, drainage divide Carpathian landscape; *10*, Chernogorye landscape; *11*, Rakhov mountain landscape.

chishcha, remain valid. The morphological classification of mountain landscapes must be to some extent two-dimensional; it must account for the vertical differentiation corresponding to the elevational horizons, and for the horizontal differentiation arising from the diversity of forms and elements of relief and a great variety of bedrock types.

Some idea of the interrelationships between the horizontal and vertical classification of mountain landscapes can be obtained from figure 34 which shows a section of the northern slope of the Great Caucasus. The diagram illustrates clearly the fragmentary character of many elevational strata and their dependence on the orographic structure and, particularly, on slope exposure. Thus certain sections of the dark-conifer forest stratum lie along the wettest valleys and the adjacent mid-mountain slopes. The distribution of pine forests represents a typical example of the effect produced by the barrier shadow.

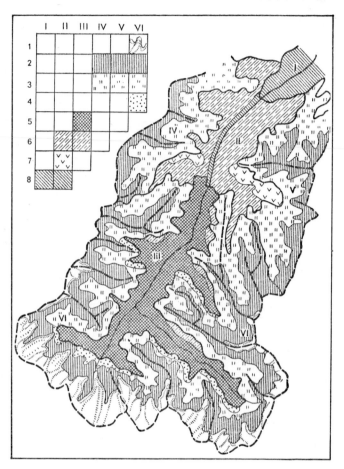

Fig. 34 Landscape chart of the Teverda River basin (after A. G. Isachenko and I. I. Tunadzhanova)

Mid-mountain landscapes in longitudinal depressions and lateral valleys: I, lower Teverdin (on sandy-clayey Jurassic deposits); II, central Teverdin (on strongly dislocated Palaeozoic sediments); III, upper Teverdin (on Pre-Cambrian crystalline rocks). *High-mountain landscapes on fold-fault ranges*: IV, Teverdin-Aksaut, and V, Teverdin-Daut (on strongly dislocated Palaeozoic sediments); VI, Dombay (on Pre-Cambrian crystalline rocks). *High-mountain strata*: 1, glacial-nival; 2, alpine rock rubble; 3, alpine meadow; 4, subalpine with sparse forest; 5, mountain tayga (dark-conifer forest); 6, pine-forest belt; 7, mountain-steppe and alpine xerophytes; 8, broad-leaved forest belt. Landscape boundaries are indicated by a thick, continuous line.

These forests dominate the driest part of the basin and, as the altitude rises, they occur exclusively on southern slopes. Fragments of the mountain-steppe stratum are firmly associated with the driest sections of the southern slopes. The subalpine scrub forest, on the other hand, occurs, also in fragments, only on the northern slopes. It is obvious that every landscape may include fragments of different strata, each of which exhibits a characteristic pattern.

A brief review of the geochemistry of landscape

Landscape geochemistry is a relatively new area of landscape science, the importance of which is gradually increasing. The foundations of this discipline were laid by B. B. Polynov who showed that the migration of chemical elements determines many important properties of landscape and hypothesized that a time would come when the study of the migration of chemical elements would become a cornerstone of landscape science.

The current 'state of the art' in landscape geochemistry has been examined in detail in a monograph by A. I. Perelman.[23] The same monograph provides a geochemical description of the principal landscape types. Accordingly we shall restrict ourselves here to a brief discussion of the importance of landscape geochemistry for the general theory of landscape science and its relationship to certain other areas and problems of landscape science.

Geochemical investigation of landscape provides one of the principal means by which the mechanism of interaction between its components and its morphological units can be studied, since such investigations are concerned essentially with the exchange of matter and energy. It follows to a large degree that the geochemical method serves as a key to our knowledge of landscape structure, in the broadest sense of the phrase.

It is necessary to distinguish the two major aspects or the two principal tasks of landscape geochemistry. The first is the study of the geochemical interaction among the components of a landscape, i.e. the cycle of elements between the vegetation cover, the biological world, soils, bedrock, water and atmosphere. A particularly important function in the exchange of matter, as well as in the transformation of energy within a landscape, belongs to the biogenic cycle. Our knowledge of the migration of chemical elements in a landscape rests on a continuous analysis of the chemical composition of rocks, weathering mantle, surface- and groundwater, soils and organisms; our attention is concentrated primarily on elements which migrate the most and which accumulate under certain physical-geographic conditions. It is natural that the study of the biogenic cycle and the migration of chemical elements should begin with individual facies (i.e. in elementary landscapes, in biogeocenoses). One example of such an investigation is provided in the work of L. E. Rodin and N. I. Bazilyevich which is concerned with the cycle of matter in certain biogeocenosis in south-western Turkmenia.[24] Facies, however, are not isolated systems with closed cycles. An exchange cycle in a given facia includes not only the material from its constituent rocks and the air column above its surface; in addition, it includes material brought from outside, primarily from adjacent facies located higher along the slope. Some material, on the other hand, may derive from a given facia, yet is transported beyond its boundaries with surface- and groundwater and atmospheric currents, through creep, subsidence, etc.

Facies which sequentially replace one another along a topographic profile form a regular geochemical system, the so-called *geochemical association*. B. B. Polynov distinguished three major elements in such systems: eluvial, superaqual and subaqual elementary landscapes (discussed earlier). A. I. Perelman designates the eluvial elementary landscape as autonomous and the two remaining types as subordinate, on the grounds of a specific geochemical relationship which exists between them. The first plays a major role in a geochemical association, since it largely predetermines the development of the subordinate facies distributed over the slopes, terraces, etc., receiving nothing from the other facies in return. Thus a geochemical association is 'a set of autonomous and subordinate elementary landscapes characteristic for each geochemical landscape . . . A geochemical association also constitutes a specific type of exchange of matter and energy among the autonomous, superaqual and subaqual landscapes characteristic of every geochemical landscape.'[25]

The investigation of the geochemical association of facies is directly related to the investigation of landscape morphology, which represents the second and a more important part of landscape geochemistry. As noted by A. I. Perelman, the migration of chemical elements binds the nature of drainage divides, slopes, terraces, floodplains and stream channels into a unit whole—a unique landscape.

Despite its obvious importance, the study of the migration of chemical elements is, of course, not the only way by which to establish the structure of a landscape. Morphological constituents of a landscape are not only associated via the downward migration of chemical elements. The lower elements of an associated series, e.g. drainage basins, floodplains, etc., affect the nature of watershed facies via local atmospheric circulation, transport of water vapour and dust, spread of pollens, distribution of animal life, etc. For this reason the idea of autonomous facies is open to question and, as Perelman himself recognizes, his choice of terminology is not entirely correct.

Landscape geochemistry is a borderline area of science which connects physical geography and geochemistry, and this connection has a reversible character. The analysis of geochemical processes is needed for the study of landscape and, vice versa, the knowledge of landscape is involved in the study of geochemical processes.

The capacity of chemical elements for migration, i.e. their mobility, is determined primarily by their internal properties: the structure of the electron shell around the atom. Accordingly chlorine is more mobile, for example, than calcium and calcium is more mobile than iron. Depending on physical-geographic conditions, however, the mobility of a given element may vary over a wide range, and the landscape itself creates the determining condition for element migration. In some landscapes Cl^- and Na^+ are the most mobile, in others Ca^{++}, and in still others Fe^{++}, etc.

The factors which directly affect the migration of chemical elements in the geographic envelope are those components of the envelope which always act in conjunction to determine the composition of active migrant elements, and the rate, direction and other characteristics of geochemical processes.

Climate determines the input of energy and moisture into a landscape. Solar energy, transformed by organisms, constitutes the most important source of energy in geochemical processes. Temperature conditions affect the rate of chemical reaction.

Almost all migration processes occur in water solutions; without moisture, chemical weathering is impossible. The character of geochemical processes is strongly dependent on the type of available water supplies, their physical-chemical properties, and their movement, and these in turn depend on climate, the organic world, topography and other landscape components. Thus the solubility of minerals and the redox ratio depend on the pH and oxidation-reduction characteristics of natural waters; in desert waters, for example, oxidation processes predominate, and in tundra and swamp waters reduction processes predominate. An important factor in the migration of chemical elements is the ascending or descending direction of motion of solutions in the soil body. Surface and underground runoff determines the redistribution of chemical elements in a landscape; it also determines the rate at which elements are eliminated from the landscape.

Rocks are the principal source of elements which enter into migration. An important factor here is not so much the overall supply of a particular element but the form in which it is found in rocks and also

172 *Landscape Science*

those properties of rocks which determine the mobility of the given element. Even very mobile elements, such as sodium, when included in rocks that are not easily eroded will exhibit, in practice, low mobility. The presence in rocks of soluble salts, on the other hand, sharply increases the mobility of the corresponding element. The conditions of occurrence also indirectly affect the mobility of elements, i.e. they affect the rate and the direction of flow of surface- and groundwater. The insufficiency or excess of mobile, bio-degradable forms of many chemical elements produces the so-called biogeo-chemical anomalies which impair various functions of living organisms and retard their normal development. Among such elements, the inadequate supplies of which in many landscapes cause anomalies of this kind, are O, N, P, Ca, Mg, K, Cu, Co, F, Mo, Zn, Mn and others; elements often supplied in excessive quantities include Cl, Na, S, Cu, Fe, F and others.

Organisms, V. I. Vyernadski's 'living matter', play a very important role in geochemical processes. Firstly, organisms bind solar energy during photosynthesis and convert it into the potential and kinetic energy of geochemical processes; secondly, organisms introduce nearly all chemical elements into the biogenic cycle, redistribute, grade and concentrate them, and in this way they alter the composition and structure of each of the three geospheres of the geographic envelope.

Organisms synthesize new (organic) substances from solar energy and from the mineral elements they absorb. The number of organic substances many times exceeds the number of known mineral compounds. The destruction of an organic substance is followed by the liberation of energy employed in its creation. But the biogenic cycle is not closed. Some substances accumulated by organisms may remain outside the cycle, forming rock accumulations and thus conserving the solar energy now converted into potential chemical energy. At the same time organic substances are frequently transported away from the given landscape and do not participate in its cycle. Living organisms play an important role in the oxygen, carbon, nitrogen, phosphorus, sulphur, iodine and many other chemical element cycles.

An important consequence of the destruction of living matter is the enrichment of water solutions in a landscape with carbon-dioxide and organic acids, which substantially increases the chemical activity of natural water supplies and their solvent (e.g. oxidizing and hydrolis-ing, etc.) capacity. Thus most of the chemical reactions in a landscape are due to the direct or indirect participation of living organisms.

The character and intensity of geochemical processes depend

directly on the mass of living material, its annual productivity, the ecological and biological properties of organisms, their capacity for selective assimilation of certain chemical elements, and other factors. All these properties of the organic world change with every landscape. Each landscape comprises a specific set of biocenoses, which in turn determines a particular type of chemical element migration. For example, in forest landscapes, where the largest mass of living matter is manufactured, water supplies are rich in organic material, and therefore the cycle is very intensive and includes the maximum number of chemical elements. In desert landscapes, with their low biological activity, the biogenic cycle is weakened and includes relatively few elements.

Finally, the geochemical significance of topography must be considered. Relief determines the direction of water flow and therefore the intensity with which chemical elements are removed from a landscape, mainly in a suspended state, and their redistribution within it. The presence of eluvial, superaqual and subaqual localities, and corresponding facies, is determined by relief. The intensity of drainage, i.e. the redox conditions, also depends on relief. Flat, lowland relief favours the location of stagnant water catchments which, if excessive moisture is supplied by the climate, leads to an insufficiency of free oxygen in the water, producing a reducing environment. Where relief is dissected, there is a rapid runoff, water is rich in free oxygen and oxidation processes predominate.

Dissected relief intensifies geochemical contrasts between the eluvial, superaqual and subaqual facies; flat relief, on the other hand, weakens these contrasts. Thus in the low-lying, poorly drained, tayga-plain landscapes, even the eluvial localities have many features of superaqual facies: shallow watertable, swamps, marked free oxygen deficit, strongly acidic water solutions and a low productivity of organic material.

Thus the character of the migration of chemical elements in various parts of the geographic envelope is determined jointly by all the components of the landscape, i.e. the landscape as such. The geochemical differentiation of landscape is subject to general zonal and azonal laws. The zonality of geochemical processes has already been discussed in chapter 2. Azonality manifests itself no less in the migration of elements, via the macrorelief and the geological base and via the effect of longitudinal-climatic factors. It is worth noting, however, that the position of a landscape with respect to the ocean and the movement of marine air masses is reflected in its geochemical characteristic not only through climate but also through the introduction into the landscape with the atmospheric rainfall of certain 'marine' ions, especially Cl^- and Na^+. In coastal landscapes, atmospheric rain contains up to 100 mg/l chlorine, whereas its content in intra-continental areas falls to 3 to 2 mg/l.

The relationship between zonal and azonal factors manifests itself in many geochemical indices and, in particular, in the intensity of 'chemical denudation', which A. I. Perelman considers an important constant. The magnitude of chemical denudation, i.e. the annual drop in ground level due to the elimination of soluble substances, is measured in microns, yet depending on the relationship between zonal and azonal factors, this fluctuates over a wide range. According to G. A. Maksimovich,[26] the steppe zone has the highest chemical denudation index (16 μ); close to it are the forest and the forest-steppe zone (13 μ). In the forest zone runoff is most intensive, while in the steppe zone runoff diminishes but the mineralization of water increases. The minimum index of chemical denudation (4 μ) charac-terizes the tundra, the waters of which are poorly mineralized and the runoff relatively negligible. Depending, however, on azonal factors (i.e. the character and solubility of the bedrock, absolute altitude and the topography) even within a single zone very strong contrasts in the transportation of matter are found. Thus in the Neva River basin the annual chemical denudation is only 4 μ, whilst in certain landscapes of the Kame basin, surrounded by highly soluble rocks, it reaches 145 μ. Mountain landscapes are characterized by more intensive chemical transportation than the lowland landscapes.

Discussion of the relevant facts indicates clearly that the geo-chemical differentiation within a landscape is entirely consistent with the principles of landscape morphology. Eluvial, superaqual and subaqual elementary landscapes are precisely those units of intra-landscape classification with which landscape morphology is con-cerned. The idea of geochemical association is closely bound up with the ideas of the genetic and ecological association between the morphological units of landscape, already discussed by L. G. Ram-yenski.

It must, therefore, be concluded that there is no difference between the geochemical landscape and the geographic landscape, except that the former concerns itself with the analysis of the migration of chemical elements and compounds. The criteria for the differentiation of geochemical and geographic landscapes are identical and there are no grounds on which could be postulated a lack of territorial coinci-dence between them. Differences can arise only in subsequent group-ing of landscapes into taxonomic categories. The classification of landscape on geochemical grounds will sometimes differ in detail from the general-geographic classification; in both cases, however, the primary elements of classification are the same territorial units—concrete landscapes.[27]

The water-heat conditions
and the seasonal dynamics of a landscape

Geographic processes in a landscape exhibit a characteristic rhythm. Best studied are the seasonal and diurnal rhythms, which differ in frequency and the causes of which are well known. There are, in addition, rhythms extending over many years, the nature of which is as yet obscure. The study of seasonal dynamics is an important part of research into landscape structure. The seasonal rhythm is an important taxonomic index as well. For example, a distinctive feature of landscapes in wet equatorial-forest zones is the continuum of all its processes in the course of a year, while temperate-zone landscapes are characterized by distinct seasonality which manifests itself in many diverse ways. The monsoon landscapes show strong contrasts in seasonal rhythm. During the summer the abundance of moisture at relatively high temperatures fosters intensive growth of the organic world and a high intensity of geochemical processes. During the winter, owing to low temperatures and the deep freezing of the soil, many natural processes cease or reduce in intensity. The mediterranean landscapes, landscapes of the western (European) and central-Siberian tayga, and the northern and southern deserts of central Asia exhibit specific seasonal rhythm.

Seasonal rhythm does not consist of a simple repetition of the same phenomena. No cycle, or rhythm, in the life of a landscape is closed; after each the landscape is irreversibly changed, and these changes accumulate year by year. A certain quantity of mineral and organic substances is eliminated, erosive dissection increases, another layer is added to the peat horizon in swamps, etc. Thus each successive cycle, diurnal, annual and any other, begins on a new, slightly altered foundation. It follows that seasonal rhythm is an inherent constituent in the progressive, gradual development of a landscape.

The study of seasonal phenomena is the subject of phenology. It is true that, until recently, phenology was restricted essentially to the study of seasonal phenomena among living organisms and was regarded as a service discipline for biology or agro-biology. The seasonal dynamics of natural processes derive from changes in heat and water conditions and its investigation should begin from the joint consideration of these factors. Despite this, even today heat and water are studied independently of one another on the basis of meteorological and hydrological observations, and consist mainly of the empirical determination of biophenological phenomena.

This dichotomy can only be overcome by approaching seasonal dynamics from the point of view of landscape. S. V. Kalyesnik observed correctly that phenology should be regarded as a geographic discipline and as a science concerned with the seasonal dynamics of a landscape. Such landscape phenology has not yet been created. Its task would consist primarily of a comprehensive analysis of the energy potential and the water conditions in a landscape, i.e. the seasonal changes in facies, urochishcha and the entire landscape. The variables include radiation and advection heat supplies, losses by evaporation from the surface and from the plants and snow, losses by transpiration, photosynthesis and soil formation, and losses by turbulent heat exchange with the lowest layers of the atmosphere, and in the thawing of snow and the melting of ice. Investigations of the water budget are concerned with the quantitative determination of the dynamics of moisture intake and outflow, its resources in a landscape, its supplies with rainfall, condensation in the soil, distribution over different facies and urochishcha, penetration into the soil, surface and ground runoff, capillary rise in the soil, evaporation and transpiration, etc. The study of the dynamics of the snow cover are important for many kinds of landscape.

The next problem is the explanation of the functional bond between heat and moisture dynamics and other processes occurring in a landscape, in particular, the budget of mineral and organic materials, the productivity of organic matter, its accumulation and seasonal losses, the biogenic cycle, the seasonal transportation of salts through the soil, chemical and mechanical denudation and the phenophase* development of biocenoses.

There is not, as yet, sufficient data on this type of dynamics for even a single landscape. There exist only approximate data obtained mainly by indirect calculation, which make it possible to compare the principal features of the heat and water balance in the horizontal direction only. For a more complete and more accurate quantitative analysis of geographic processes we need field investigations extending over several years. A classical example of such investigations is the twelve-year study, during 1892–1904, by G. N. Vysotski in the Vyeliko-Anadol, which has remained unrivalled as a comprehensive account of the phenomena involved.

In recent years geographic investigations by means of observation

---

* 'Phenophase' refers to the annual or semi-annual period or cycle of development of a plant or plant community.

stations have developed somewhat, but in the main they remain relatively specialized and are not concerned with the full set of phenomena which interests the geographer. Thus the Geographic Institute of the U.S.S.R. Academy of Sciences is conducting a comparative field investigation of the heat balance over the beech-aspen forest and the grasslands near Moscow. These investigations embrace the principal elements of the heat balance: radiation budget, heat losses including transpiration in total evaporation, and the heat exchange between the soil with its vegetation cover and the ground layer of the atmosphere.

Our knowledge is especially deficient with regard to energy relationships in soil-biological processes. According to V. P. Volobuyev,[28] the total volume of energy participating in soil-forming processes or, more correctly, soil-biological processes, is subject to the following losses: (1) in the physical destruction of soil-forming rocks; (2) in the chemical decomposition of minerals in such rocks; (3) in energy accumulation in the humus; (4) in biological reactions converting organic and mineral substances; (5) in evaporation from the surface of the soil and the vegetation; (6) in transpiration; (7) in the processes of mechanical migration of salts and fine particles in the soil; (8) heat exchange losses in the soil-atmosphere system, the latter being often close to zero in an annual budget.

By means of only very approximate calculations, V. P. Volobuyev found that the total annual loss of solar energy in these processes amounts to 2 to 5 kcal/cm² in the tundra and the desert, 10 to 40 kcal/cm² in the temperate-forest and steppe zones and 60 to 70 kcal/cm² in wet tropical regions. Most of this energy (between 95 and 99·5 per cent) is lost by transpiration and evaporation and only 1 per cent on the average (up to 5 per cent in wet tropical zones) by biological processes; a negligibly small fraction (approximately 1 per cent) is lost by the processes of mineral decomposition.

Great interest is attached to changes in the relationships between the individual elements of an energy budget in the course of a year. Such data, however, are almost entirely lacking. V. P. Volobuyev quotes one example, i.e. for the dark-chestnut soils of the so-called steppe plateau in Azerbaijan, which is also based on indirect calculations. In this area 25·3 kcal/cm² of solar energy is lost in the course of one year by soil-biological process, including 24·8 kcal/cm² in transpiration and evaporation and approximately 0·5 kcal/cm² in photosynthesis (1·8 per cent of the total loss). During the growth period, up to 5 per cent is lost in the synthesis of organic material (during May to August the monthly loss amounts to approximately 0·1 kcal/cm²; the maximum falls in June). Starting with February, some of the energy lost in the production of the organic mass is liberated with the latter's decay (a maximum of approximately 0·1 kcal/cm², reached in October and November). Heat losses by evaporation and transpiration combined are at their highest in April (4·6 kcal/cm²) and May (5·2 kcal/cm²). The available data on the water balance are usually limited to the relationship between rainfall, runoff and evaporation. M. I. Lvovich characterized in some detail the water balance in various landscape zones discussed earlier in chapter 2 (see table 1). There also exist studies of the water balance of individual ugodya,* both forested and unforested, which illustrate the relationships between rainfall, surface and ground runoff, transpiration and evaporation from the ground surface in forests and also from the tree-crown surface. These studies, as a rule, are concerned with mean annual values and do not reflect the dynamics of moisture supplies in the landscape or the changes in the relationships between individual components in the course of a year. A. G. Isachenko studied the seasonal dynamics of moisture in a landscape for thirty-three points distributed over various zones of the Russian plain.[29] Some typical results are illustrated in figure 35.

The entire annual cycle of moisture exchange may be divided into two major periods: the period with a positive moisture balance when

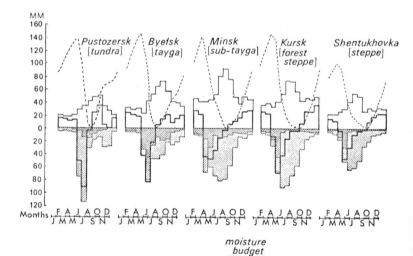

Fig. 35   Moisture dynamics in landscapes belonging to different zones

* Agricultural land.

the amount of rainfall exceeds losses in evaporation and runoff, a period with a negative balance when the loss by evaporation and run-off exceeds rainfall. During the former period, moisture is accumulated in the landscape and supplies are stored; during the latter period these supplies are exhausted. The sharpest transition in the water budget occurs when air temperatures change from negative to positive, i.e. at the beginning of spring. At this time moisture supplies in the landscape will have reached a maximum, due to snow and the accumulation of liquid resources in the soil and the subsoil, but as temperatures enter the positive range, these supplies are gradually depleted due, in the early stages, to the intensive run-off of melting waters. At the same time the evaporation curve rises quickly. The water budget becomes negative. The termination of the spring flood-time marks the end of spring and the beginning of summer.

Summer is characterized by the most intensive moisture exchange. During summer most of the rainfall is precipitated and utilized mainly in transpiration; the runoff is greatly reduced, and in some landscapes disappears altogether. Owing to more intensive transpiration, the supplies of moisture in the soil body are progressively depleted and a moment comes when these supplies are at a minimum. Such minimum levels may be designated as *passive*, since these waters no longer participate in the exchange process. Thus the moment of maximum dryness (or minimum moisture) occurs in a landscape when the water supplies accumulated during the autumn-winter rains, of which the proportion participating in the exchange may be called *active*, are completely exhausted. Since moisture accumulation takes time, the period of minimum real wetness does not coincide with the dryest month in the climatic sense, i.e. when the rainfall-evaporation ratio is at its annual minimum. The driest time of the year, with respect to real water supplies, occurs some one to three months later, and the drier the climate, the greater the delay.

The period when moisture supplies in the landscape are at a mini-mum, i.e. when the active water supply is at a zero level, may be considered mid-summer. During the first half of the summer the water budget is negative; growth is most intensive at this time due not only to concurrent rainfall but also to the autumn-winter accumulation. During the second half of the summer, water losses are reduced and are fully compensated by rainfall; the water budget changes from negative to positive and accumulation is resumed. The higher the rate of drying during the growth period, the more intensive is the absorption of moisture at this time and, for this reason, a high pro-portion of the active water supplies in the southern half of the Russian plain is made up of liquid supplies absorbed by the soil, whereas in the tayga and the tundra the active supplies are mainly due to snow accumulation, all or most of which is converted into runoff.

The transition from summer to autumn is very gradual; in the tundra and the tayga, autumn is characterized by a secondary runoff maximum, while in the steppe zone the second half of the summer constitutes the beginning of autumn. Winter is characterized by the accumulation of moisture in a solid phase and by a sharp reduction in evaporation and runoff. It begins, as far as the water supply conditions are concerned, approximately one month after the mean annual temperature descends below freezing.

Attention is now directed to the principal features of the seasonal dynamics of the water supplies in the various zones of the Russian plain. In the tundra a negative water budget exists only during two to three months (i.e. May to July), while moisture continues to accumulate during the remaining months. In all seasons the amount of runoff exceeds evaporation. Active moisture supplies, almost entirely in the form of snow, reach approximately 140 mm during April; during May to June these supplies are converted into runoff, which consists not only of the entire snow supplies but also of a proportion of concurrent rainfall (approximately 30 to 40 mm). Accordingly, the autumn to winter moisture accumulation does not play a role in supporting the growth of the vegetation cover. The summer (July to September) is characterized by a distinct low-water period, but more water is still received than evaporated; during July (the first half of the summer) the water budget is still negative in certain localities, but this is due to the continuing runoff of the residue from melted snow. During the autumn (October to November) the evaporation is slight and a secondary runoff maximum occurs; during the winter (December to April) the moisture exchange is substantially reduced, the runoff is at a minimum and evaporation is negligible.

In the tayga the moisture budget is also positive during a large part of the year, except between April and June. The maximum active water supplies amount to 150 to 200 mm (usually during the middle or the end of March), of which some 120 to 150 mm are contained in the snow cover and the rest constitutes liquid supplies in the soil and groundwater. During the spring (April to May) most of these supplies (80 to 100 per cent) are depleted with the spring floods, but a certain excess (approximately 30 mm) is retained in the southern tayga to supplement the low-water runoff and possibly, sometimes, to moisten the soil and support transpiration. During the spring the evaporation is relatively intensive; in May it approximates runoff in quantity. During the summer (June to September) the evaporation, principally in the form of transpiration, is at its most intensive and somewhat exceeds the amount of runoff although it is, as a rule, fully compensated by rainfall supplies. During autumn (October to November) the evaporation is sharply reduced while the runoff increases (a secondary maximum). Winter, with minimum amounts of runoff and evaporation, lasts on the average from December to March.

In the subtayga zone (mixed forests) the period with a negative water budget is increased to four months, i.e. from March to June. Active water supplies reach a maximum level at the end of February or early March and remain roughly on the same level as in the tayga, but the proportion of snow is lower and fluctuates considerably depending on the temperature conditions during the winter; in the east it still amounts to some 100 to 130 mm, but in the west it falls to 20 to 40 mm. The spring floods (March to April in the west, April to May in the east) carry away approximately 70 to 80 per cent of rainfall supplies accumulated during the autumn-winter period, whilst the remainder (30 to 40 mm) is partly lost by evaporation and partly retained in the soil. Summer, autumn and winter are virtually the same as in the tayga, but the excess of evapotranspiration over runoff is much more pronounced. The vegetation's demand for moisture during the first half of the summer (May to June) is not always satisfied by the concurrent rainfall.

The forest-steppe zone is characterized by an even longer period with a negative moisture budget (March to August or April to August), a sharply reduced runoff and considerable evaporation. The accumulation of active water supplies is less intensive; the maximum (100 to 160 mm) coincides with the end of February or the beginning of March. As in the previous zones, the proportion of water supplies obtained from melting snow varies from approximately 120 mm in the east to some 20 mm and less in the west. In the west the spring begins earlier than in the east. Spring flooding (March to April in the west, April to May in the east) absorbs 50 to 70 per cent of the accumulated moisture supplies, while 40 to 50 mm supplement the supplies used in transpiration. During the first half of the summer (May to July in the northern forest-steppe, May to August in the southern) the rainfall is usually insufficient to cover the losses in transpiration and evaporation. Only at the very end of summer (September and, in parts, August) when transpiration begins to decrease does the water budget become positive. The autumn generally extends over two months (October to November) and the winter over three months (December to February). During all seasons, except spring, the runoff is fairly low and the autumn maximum does not occur, since the dry soil absorbs the major proportion of unevaporated rainfall.

In the steppe zone the atmospheric rainfall for one half of the year

(March to August) does not compensate for the loss of moisture, i.e.
the water budget of the landscape is negative. The runoff is very small;
in the southern steppe subzone it occurs only during the spring. The
active moisture supply reaches a maximum level of 100 mm during
February or March, and consists essentially of the moisture absorbed
during the autumn by the soil, the moisture of which was depleted
during the growth period. In the east the thawing of the snow cover
plays an important role in wetting the soil. In the west the supplies of
water in the snow cover are low during the spring thaw, but during the
winter, owing to periodic thaws, certain amounts of water enter the
soil. The spring flooding in the steppe lasts a short time (in the eastern
part it occurs in April, in the west mainly during March) and absorbs
less than 30 to 50 per cent of the moisture supplies accumulated
during the autumn-winter period. As the spring approaches, the soils
contain 50 to 70 mm of moisture, which may be subsequently lost by
evaporation. From April to August the evaporation (in a broad sense,
i.e. including transpiration) exceeds rainfall and the deficiency is
compensated by the supplies held in the soil; since, however, the total
supplies are rather low, the amount evaporated, having reached a
maximum during May, begins to fall sharply from June onwards. The
budget becomes positive only during September. In fact the second
half of the summer in the steppe actually constitutes early autumn
(September to November). During this period the evaporation is low
and the runoff negligible, often even smaller than during the summer,
since all excess moisture not subject to evaporation is absorbed by the
dry soil. During the winter (December to February and, in the east,
also in March) there is virtually no runoff.

Each individual morphological unit in a landscape has its characteristic seasonal rhythm associated mainly with the local climate or
microclimate. Intra-landscape differences in the seasonal rhythm
have so far been poorly studied. One interesting example is provided
by the phenological investigation of D. F. Tumanova carried out at
the Leningrad State University field station in the north-western Lake
Ladoga region (within the area shown in figure 30)[30] during which the
following facies were studied:
1. Selga crest, with sparse lichen pine forest, 1800 m south-west of
the Lake Ladoga shore ('Suurian selga').
2. Selga crest, with sparse lichen and bilberry-pine forest, 100 m
from the lake shoreline ('Lake Ladoga selga').
3. The lower part of the south-western slope of 'Suurian selga', with
small-leaved deciduous and conifer mountain-sorrel forest, 1800 m
south-west of Lake Ladoga (outside its immediate influence).
4. The lower part of the south-western slope of the 'Lake Ladoga

selga', with conifer small-leaved deciduous mountain-sorrel forest, 700 m south-west of Lake Ladoga (outside its immediate influence).
5. The western foot of the selga, with mountain-sorrel spruce forest, 230 m south-west of Lake Ladoga, open in the direction of the lake.
6. The northern foot of the 'Lake Ladoga selga', with conifer small-leaved deciduous bilberry forest, along the lake's shore (1 m from the shoreline).
7. Inter-selga hollow, with timothy grass and clover meadow, near Lake Suuri (1000 m from Lake Ladoga and approximately 200 m from Lake Suuri).
8. Inter-selga hollow, without lakes, with spikelet-bent-grass meadows, 2550 m from Lake Ladoga.

Some indices of seasonal rhythm for each of these facies are given in table 3 (area numbers correspond to the facies above).

There are obvious differences in the seasonal rhythm of individual facies during the year, including the winter period. Thus a stable snow cover is established earlier in hollows, where it becomes thicker and more dense. The most striking differences, however, occur at the beginning of spring. (The onset of spring coincides with the beginning of the thaw.) The onset of late spring coincides with the flowering of the liverwort (*Hepatica nobilis*) and the grey alder (*Alnus incana*). The onset of early summer coincides with the end of the flowering of the bird cherry (*Padus racemosa*) and the mountain ash (*Sorbus aucuparia*) and the flowering of the star-flower (*Trientalis europaea*). The onset of late summer coincides with maturing berries (*Vaccinium myrtillus*); the early autumn period coincides with the beginning of November, late autumn with the end of November, early winter with the establishment of a stable snow cover, and late winter with the deep freezing of the soil. The earliest onset of spring and its most rapid passage characterizes the crests of the selga located at some distance from Lake Ladoga. (The Lake Ladoga selgas are slightly delayed owing to the cooler conditions.) The most delayed and the longest phases of seasonal dynamics occur along the Lake Ladoga shore and on foothill facies with closed spruce forest where the snow is retained longer than in other localities, and all the spring-summer phenological phenomena are delayed by one to two weeks by comparison with the hollows, which are cultivated, and the crests of the selgas. During the summer the crests and upper slopes of the selgas and the dry hollows have the highest temperatures, while the Lake Ladoga shoreline has the lowest temperatures. The foothill areas of the selga ridges, protected from the influence of the cold lake, have temperatures in between these extremes. Autumn frosts begin in the dry hollows (i.e. without lakes or forests). Autumn also comes early to the crests of the selga,

Table 3

No. of area	1	2	3	4	5	6	7	8
Snow cover (cm) 5 December 1955	33	32	32	36	25	35	46	47
Snow cover (cm) 28 March 1956	60	70	73	75	60	78	85	85
Density of snow cover (kg/m³) 28 March 1956	0·26	0·24	0·27	0·22	0·27	0·25	0·34	0·30
Beginning of late-spring period (1956)	5 May	7 May	7 May	6 May	9 May	10 May	—	2 May
Beginning of early-summer period (1956)	17 June	19 June	19 June	19 June	21 June	20 June	—	13 June
Beginning of late-summer period (1956)	10 July	10 July	12 July	21 July	23 July	26 July	—	—
Mean air-temperature between 27 June and 13 August (°C) (1955 and 1957)	20·6	18·9	19·8	18·1	17·1	17·6	19·5	20·1

caused apparently by the extremely variable rainfall and intensive drying out of the soil towards the end of summer. Autumn sets in last of all along the lower slopes of the selga. It is most delayed in the spruce forest at the foot.

The analysis of seasonal rhythm is very important in the classification of the morphological units of landscape. Facies and urochishcha which outwardly appear identical, in fact turn out to be very different when their seasonal dynamics are taken into account.

The development of a landscape

Even the early protagonists of Russian landscape science, including L. S. Berg, G. F. Morozov and B. B. Polynov, regarded landscape as a complex material system undergoing continuous development. 'We can only comprehend a given landscape', wrote L. S. Berg, 'if we know how it originated and what form it assumed during its life-span.'[31]

All current definitions of landscape emphasize its genetic singularity and homogeneity. Each landscape is treated as a single developing entity with a specific origin, characteristic growth and history of development.

Relatively little, however, is known about the activating forces and the laws of landscape development, and some geographers consider the possibility of applying the notions of 'genesis' and 'development' to landscape to be somewhat doubtful.

L. S. Berg was among the first geographers to pose the question regarding the forms of landscape development. He distinguished two types of change in landscape: *reversible* and *irreversible*. The first include the seasonal changes which 'do not, essentially, introduce anything new into the existing order of things', as well as catastrophic changes (e.g. earth tremors, large fires, rodent infestations, etc.), following which the landscape 'returns to a condition resembling the pre-catastrophic'.

Following the irreversible or progressive changes 'there is no return to the original state: changes occur in a single, well-defined direction'.[32] Among the causes of progressive changes in landscape, L. S. Berg listed the effect of climate, geological factors (e.g. upthrusts and downthrusts, marine transgressions and regressions), as well as the activity of biological organisms and man. L. S. Berg's important contribution to the study of landscape development was the idea that changes in landscape may be due to both external and internal causes (e.g. the interrelationships between plant communities and the environment). He did not, however develop this idea in any detail.

Until recently geographers usually attributed landscape development to some major external factor (e.g. changes in solar activity, tectonic movements, displacement of the earth's poles) or to changes in some component generally regarded as dominant. This role was usually assigned to that component which reflected most directly the effect of external-terrestrial factors (i.e. climate and the geological base). Some geographers treated the dynamics of the vegetation cover and its interaction with the geographic environment as the major factor in the development of geocomplexes.

The fact that landscapes on earth continuously undergo major irreversible changes under the action of external cosmic or tectonic forces is a well-known fact. A more important question is: Is landscape subject to *self-development*, i.e. does it change progressively as a consequence of *internal contradictions*? In accordance with the principles of dialectic materialism, every material system develops in consequence of a competitive struggle among opposing forces which is inherent in the system. The resolution of internal contradictions is the energizing force in development.*

It is possible to find interesting ideas about the self-development of natural complexes even in classical texts of Russian geography. Thus V. V. Dokuchayev has shown that the development of lakes proceeds undeviatingly in one specific direction even if all external conditions remain constant. V. N. Sukachev[33] showed clearly the nature of the self-developmental processes in a geocomplex, using as an example the simplest complex—biogeocenosis. Sukachev found that there exist contradictions among the components of a biogeocenosis which act as its developmental force. Thus the biological activity of vegetation gives rise to contradictions between it and the environment. Vegetation continuously alters the environmental conditions and, at the same time, adapts to the changes it has wrought; there is a persistent tendency towards an equilibrium between vegetation and other components, yet this equilibrium is continuously disrupted by the vegetation itself. Such contradictory internal relationships also exist between the atmosphere and the geocomplex, and between the lithosphere and the geocomplex, etc.

New elements, new facies (biogeocenoses) continuously develop and augment the landscape, gradually altering its character until the

---

* One of the three basic laws of dialectics is: 'Internal contradictions are inherent in all phenomena of nature and society, all have their negative and positive sides, a past and a future, something dying away and something developing; and the "struggle" between these opposites constitutes the internal content of the process of development. "In its proper meaning, dialectics is the study of the contradiction within the very essence of things" (Lenin). The unity is conditional, temporary; the struggle absolute'. (Gould, 1967)

new form replaces the old. B. B. Polynov and L. S. Berg drew attention to the fact that every landscape includes elements of a different age. B. B. Polynov[34] distinguished in landscape *relict, conservative,* and *progressive* components. These relict components are inherited from past geological epochs and provide us with evidence of the past developmental history of the landscape. Included among them are relief forms (e.g. glacial landforms), components of the drainage network (e.g. dry riverbeds in the desert) biocenoses and soils (e.g. the steppe communities in the tayga on appropriate soils) and entire facies or urochishcha. These conservative components fully correspond to the existing conditions and determine the present-day structure of the landscape. Finally, progressive components underscore the dynamic character of the landscape and reflect the trends for its future development, providing the data for predictions about its future. An example is provided by the emergence of forest and steppe islands, the appearance of patches of thawed ground in a permafrost district, and of erosion forms of relief in morainic landscapes produced by the most recent glaciation. The gradual quantitative accumulation in a landscape of the components of a new structure eventually leads to such radical, qualitative changes that we are forced to speak of a transformation of one landscape into another.

From outside, the process of self-development is often regarded as a consequence of changes in a single component, which is viewed as having a major role. A more perceptive analysis, however, leads to a completely opposite conclusion. It turns out that a change in such a component (e.g. relief) is itself a consequence or a partial manifestation of the process developing the landscape as a whole. Thus some geographers attributed the dynamics of forest-steppe landscapes and the relationship between forest and steppe to the dominating effect of relief-forming processes and, specifically, to progressive erosion. But what determines the development of relief? The answer is: the unavoidable and regular interaction between relief and the remaining components of a landscape, even if the tectonic and other external conditions remain constant.

The western-European forest-steppe comprises a combination of geographic conditions, a structure of landscapes which greatly favours an intensive development of erosive processes, e.g. easily eroded soils, torrential rains, rapid thawing of snow, absence of forest vegetation, and intensive runoff, each factor in turn aggravated by man's activity. But a more intensive dissection of the surface subsequently alters the water supplies, the relationships between the forest and steppe communities, etc. These changes are usually regarded as resulting from the development of relief, whereas in reality they constitute a precisely opposite process in which cause and effect alternate in accordance with the laws of dialectics. The same relationships occur between the landscape and the climate, the landscape and the vegetation, etc.[35]

The internal development of a landscape proceeds relatively slowly and is very rarely observed in a pure form, since superimposed on it are the changes caused by factors external to that landscape: changes in atmospheric circulation, upthrusts and downthrusts of the earth's crust, oceanic transgressions and regressions and continental glaciations. These external influences may not only channel the normal development of a landscape in a particular direction (i.e. accelerate or retard it) but may also prevent it altogether, thus, as it were, destroying the landscape. During the Quaternary period, for example, many landscapes completely disappeared under the ice cover and new landscapes subsequently emerged in their place; others disappeared or, more correctly, were transformed into underwater landscapes, as a consequence of marine transgressions.

The gradual changes in landscapes are complicated by the various kinds of cyclic variations which impart to the developmental process an outwardly oscillating (reversible) character. Finally, most landscapes today are subject to man's interference which assumes a great variety of forms. (This is discussed in more detail in the next section.)

Thus every landscape is subject to a variety of influences acting simultaneously, and this creates serious difficulties when one attempts to determine its history or to explain the participating forces and the mechanism of its development.

The process of internal development is much more easily studied in terms of the simple geocomplexes which are often observable over a period of time, e.g. when a lake is overgrown, a rivercourse meanders, when forest encroaches on the steppe, or when gullies are produced. Processes of this type eventually result, as mentioned, in a change in the morphological structure of the landscape, but such complete restructuring obviously requires a much longer period.

Another complex problem is the determination of the age of a landscape. A new landscape emerges when new structural elements have appeared in it, such as new climatic features, biological features, soils, and other components, as well as new morphological structures. The emergence of a new landscape may be due to both external and internal factors. So far, however, we are much better acquainted with the changes due to external factors (e.g. glaciation, tectonic movements, changes in land and sea boundaries, etc.). The reason is that these changes are frequently catastrophic in character, they take place very rapidly in terms of geological time and are more clearly evident in the landscape.

The moment when land emerges from under the sea, or from under the ice cover, always signifies the appearance of new landscapes. This does not mean, however, that the age of contemporary landscapes should be assessed from the time when they first emerged from under the sea or the ice. Much of the earth's surface has not been under the sea for many millions of years and has existed as dry land since the Mesozoic, Paleozoic, or even earlier eras. This does not mean, however, that contemporary landscapes in such areas are of the Mesozoic, Paleozoic, or still older ages. Even in geologically very young areas, which only 10 000 to 15 000 years ago emerged from under the ice cover, landscapes would have emerged many times owing to macroclimatic changes. It is well known that in northern Europe landscape zones have shifted many times since the Valdai glaciation. Obviously changes in landscape zones bring about corresponding changes in landscapes.

It follows that the age of the landscape should not be confused with the age of its geological base or with the age of the area within which the landscape develops. Contemporary landscapes which originated mainly in the Cenozoic are in virtually every case much younger than the base which underlies them. An exception are the landscapes which emerged in recent times along sections of sea-bottom (e.g. along the shore of the Caspian sea, drained as a result of a fall in its level). The original landscapes still exist in such areas, enabling us to study the initial processes in landscape formation which coincided with the area's emergence from under the sea.

Two major stages in the development of a landscape in a new area can be distinguished. The first stage constitutes a gradual though relatively rapid development of those structural features of a landscape which are determined by the geographic location of a territory in the zonal-azonal system, by the local tectonic movements and by a specific 'historical heritage'. Even after catastrophic changes a certain continuity persists between the new and the old landscape. The new landscape inherits from the old the most conservative components: the geological base and relief, which constitute relict forms in the new landscape.

Originally homogeneous areas of the old sea- or lake-bottom are subjected to the effects of solar radiation, atmospheric circulation, water flow and settlement, all of which gradually differentiate and complicate its morphological structure. During this stage of development the landscape is characterized by rapid variations and exhibits typical features of a young and as yet unformed structure: incompletely formed biocenoses (in V. N. Sukachev's terminology, a state of syngenesis) poorly developed soils, poorly-dissected topography, and an undeveloped drainage network.

Eventually all these components reach a certain equilibrium with one another and with the overall zonal and azonal conditions; the landscape attains a relatively stable structure. From this time on the development enters a second, very much longer stage during which changes originate in the competitive interaction between its components. These subsequent changes may arise from a very gradual restructuring produced by the accumulation of progressive elements. In cases where the change from an old to a new landscape is due to internal development, it is extremely difficult to draw a sharp boundary between the old and the new, since the quantitative changes extend over a very long period, assuming that there has been no interference from the external factors. In practice, however, the external factors nearly always upset the natural developmental path of a landscape and constitute the major guides to the interpretation of its history.

The effect of man on a landscape, and 'cultural' landscapes

Human society has deeply altered landscapes on our planet. There are today virtually no landscapes which do not directly or indirectly reflect the economic activity of man. Many landscapes have lost their original appearance to such an extent that we now call them 'cultural' landscapes (by contrast with 'natural').*

We cannot, however, draw a sharp boundary between cultural and

* The terms 'natural landscape' and 'cultural landscape' (originally translated as 'cultured') were taken from German geography and popularized as technical jargon in American geographic circles by Sauer. (1925) The distinction of natural and cultural landscapes has been criticized severely by Hartshorne:
'The form in which Sauer states his concept of these two terms . . . would seem to indicate that they were separate components of the total "landscape", the former consisting of all the natural features of an area, the latter of all the man-made forms. Though this might be a logical division of "landscape" when understood simply as the collection of material features of an area, the suggestion that the cultural forms that man has erected could be separated from the natural base on which they are built and thus considered as any kind of a "landscape" would appear to stretch that poor word beyond recognition.' (Hartshorne, 1939, pp. 346-50)
'Many writers describe the aspect of the face of the earth which results from the presence of man as the cultural landscape—the natural landscape modified by man. Perhaps the chief objection to this is that it suggests the existence of two landscapes—a natural and a cultural one. As a matter of fact, the fundament ceases to exist and is replaced, after the arrival of man, by the cultural landscape. There is after all, only one landscape.' (James, 1934, p. 80)

natural landscape, since even the most deeply transformed landscapes remain parts of nature and develop according to natural laws and not social laws. Man cannot change natural laws, but he applies them to his advantage, spontaneously or rationally altering the direction and the rate of natural processes.

Not all components of a landscape are subject to equally intensive transformation by human society. The geological base, type of relief, and more important features of the climate remain unchanged in every cultural landscape. Man is as yet not in a position to alter the fundamental (zonal and azonal) conditions for landscape development. As a result, as soon as man's interference is terminated, the landscape tries to return to its original state. It is true that a landscape which has once been subject to man's interference is no longer fully reversible. Yet such irreversibility is also a feature of natural landscapes. Man's influence on a landscape becomes fully irreversible in cases where he knowingly or unknowingly assists the natural development process, thus accelerating the development of a landscape or liberating the hidden potential of a landscape. Irreversible, for example, are certain natural geomorphological processes induced by man's interference (e.g. gully formation, soil erosion, etc.). The potential for such processes already exists in the structure of the given landscape, but their development in natural conditions would be greatly retarded.

A permanent situation is established when species of plants or animals enter a landscape which affords them more favourable environmental conditions and from which they were previously barred by historical causes. Thus the propagation of rabbits introduced to Australia in 1859 assumed a veritably catastrophic character; the rabbits not only established themselves as an inherent element of the zoocenoses but substantially affected the plant cover and seriously damaged the grazing capacity.

Occasionally forest plantations are introduced in areas where natural forests have been felled by man, or where the natural conditions are suitable but where forests have failed to establish themselves (e.g. in the forest-steppe zone). These constitute a permanent alteration and could remain in the landscape at least for as long as the appropriate zonal-azonal conditions for its existence are preserved.

When, however, man introduces into landscape changes which could not occur naturally, the changes produced will persist only if man's interference is continued, i.e. if the necessary conditions are constantly maintained. As soon as man ceases to maintain these conditions, the landscape gradually returns to its original state. Thus forest plantations in the dry-steppe or semidesert areas, the maintenance of which requires constant irrigation, are lost as soon as water supplies are cut off. Pastures and orchards can be maintained in the desert under artificial irrigation. Yet left to themselves they rapidly disappear and the original desert landscape is re-established in their place. Drier areas of the tayga revert to swamps wherever artificial drainage networks are neglected. Changes introduced by man's temporary (or accidental) interference, e.g. through forest fires, are reversible. After a tayga fire the plant cover and eventually the animal communities are re-established.

The consequences of man's economic activities on a landscape are two-fold. Direct effects consist of changes in individual components and in the appearance of productive modifications of certain morphological units. The elements subject to such direct effects are primarily the vegetation cover, the animal world, soils, water supplies and the drainage network. There are many examples of a radical alteration of the plant and animal communities over vast areas, such as the steppe or the forest-steppe; there are also many examples of the transformation of a drainage network achieved by the construction of dams, canals, etc. We are concerned in such cases not with man's interference with individual geographic components, but with the effect of his activities on the structure, morphology and development of the landscape as a whole.

It was noted earlier that man's activities are reflected most rapidly and directly in the morphological units of landscapes. Man's productive activities embrace individual cultivated lands, i.e. morphological units of landscape brought into economic use. This results in intensified intra-landscape differentiation induced by the emergence of different kinds of modifications which diversify the landscape yet themselves remain essentially reversible. Thus within the boundaries of a single landscape there may exist simultaneously and in varying quantitative ratios morphological units which man has altered in some degree, from areas only slightly affected by economic activity (e.g. swamp areas) to radically transformed areas (e.g. pastures, orchards on reclaimed swamps, cleared forest, etc.) with many intermediate stages (e.g. productive forests, scrub, fallow land, etc.).

In addition to these direct results, man's interference produces a number of unexpected consequences associated with processes which are unavoidably outside man's control and arise or intensify due to the disruption of geographic interrelationships. These consequences are often irreversible and as a rule undesirable. As Engels wrote:

The people who, in Mesopotamia, Greece, Asia Minor, and elsewhere, destroyed the forests to obtain cultivable land, never dreamed that they were laying the basis for the present devastated condition of these countries, by removing along with the forests the collecting centres and reservoirs of moisture. When, on the southern slopes of the mountains, the Italians of the Alps used up the pine forests so carefully cherished on the northern slopes, they had no inkling that by doing so they were cutting at the roots of the dairy industry in their region; they had still less inkling that by doing so they were thereby depriving their mountain springs of water for the greater part of the year, with the effect that these would be able to pour still more furious flood torrents on the plains during the rainy seasons.[36] What did the Spanish planters in Cuba, who burned down forests on the slopes of the mountains and obtained from the ashes sufficient fertilizer for one generation of very highly profitable coffee trees, care that the tropical rainfall afterwards washed away the now unprotected upper stratum of the soil, leaving behind only bare rock?[37]

One could endlessly enumerate examples of such negative indirect effects of a one-sided destruction of the structure of a landscape. The destruction of forests, of natural grass cover, irrational agricultural practices, etc. continue to produce erosion, loss of soils, reduction in water supplies, silting of reservoirs, and the propagation of pests such as insects and rodents, etc. Over-intensive grazing leads to the spread of sandy desert and to the replacement of valuable with valueless vegetation over large areas. Irrigation is often associated with the rise of the groundwater table and salination, whilst the construction of dams and water storages may turn neighbouring areas into swamps.

In most cases these activities also intensify the morphological differentiation of a landscape owing to the appearance of certain 'unwanted' facies and urochishcha (e.g. gullies, secondary solonchaks, swamps, mobile sand, etc.). The problems of human-induced changes and the problems of re-establishment of regional facies have recently been examined in detail by the Latvian and Lithuanian geographers K. G. Raman, A. B. Basalikas, A. A. Seybutis and O. Shleynite. Their investigations have shown that the original forest facies were transformed into sparse forest, forest-scrub, mossy glades, and arable fields at different stages of cultivation. Moreover, neglected arable lands gradually reverted to a condition close to the original, passing (according to K. G. Raman) through a sequence of 're-naturalization' stages: glades, sparse forest and secondary forest. This process was repeated many times in the course of centuries, yet it is not completely reversible.

Certain irreversible processes caused or accelerated by man's activities played a characteristic role in the intensification of the morphological dissection of Baltic landscapes, as well as many others. Thus farming activities along the slopes caused the erosion of the cultivated topsoil and its redeposition along the frontal apron in the form of characteristic benches up to 2 m thick, together with a composite profile. Farming over blind creek hollows contributes to the development of swamps. Large amounts of mineral substances are introduced, leading to a change in the vegetation and in the properties of peat; a deluvial deposit is created around the edges, while in the centre, oligotrophic facies are replaced with eutrophic.* Thus man's activities disturb the natural character of the landscape and the relationships between the individual facies and the urochishcha are altered.

Man's activity, however, does not always intensify the differences within a landscape. In some cases the contrasts between individual facies and urochishcha are smoothed out by land reclamation and the improvement in fertility of unproductive soils, by irrigation, and by the enlargement of cultivated areas. Activities of this kind correspond to a higher development level of productive forces, whereas intensified fragmentation of landscapes is typical of the primitive farming activities of the past. Man's activities also extend indirectly to 'virgin' landscapes which do not participate in man's economic or other activities. The most important effect of these activities is the change in the composition of the atmosphere due to the combustion of fuels causing an increase in the content of carbon-dioxide. An important factor nowadays is the contamination of the atmosphere with radio-active isotopes liberated by nuclear explosions. Owing to atmospheric circulation, radioactive rainfall may be precipitated far away from the site of explosion, contaminating soils, water supplies and organic life in different landscapes.

The degree and the character of man's influence on landscape depends on social-historical conditions and the level to which industry, science and technology are developed. Even primitive man exerted an influence on the landscape through his hunting, fishing and primitive farming activities. These early activities did not introduce any substantial changes in the earth's landscapes. More active interference began with cattle grazing and grain cultivation, followed by cottage industry, yet even these activities did not compare in scale with those which human society introduced into landscapes during the last century, let alone the past few decades. Differences embrace not only an increased intensity of man's interference with landscape; as human society develops, man's relationship to the landscape and to the character of his interference with natural processes alter fundamentally.

Engels wrote that 'All hitherto existing modes of production have aimed merely at achieving the most immediately and directly useful

---

*'Oligotrophic' refers generally to lake-habitats characterized by 'steep or rocky shores and scanty littoral vegetation'; 'eutrophic' refers generally to lake-habitats characterized by 'gently sloping shores and a wide belt of vegetation'.

effect of labour.'[38] When man interfered with nature crudely and destructively, nature, in Engel's words, paid him back with unexpected consequences (some examples were discussed earlier).

In a socialist and communist society, the utilization of natural resources must be based on the understanding and on the correct application of the laws of nature. In practice this means a comprehensive and complex investigation of the mutual relationships between the components and the morphological units of a landscape, and among landscapes themselves. The landscape as a whole and not its individual units should be the object of human activity and transformation. The period of socialist construction has also seen examples of negative interference with landscape (e.g. through excessive deforestation, water pollution, etc.). But these examples are not part of a pattern. They are evidence of inadequate planning and could have been avoided if geographers had fulfilled their responsibility in the drafting of current and future plans of economic development.

It follows that the classification of landscapes into natural and cultural is based on an over-simplification. On the one hand, it removes, so to speak, cultural landscapes from the effects of natural laws. On the other, such a classification does not reflect the multiplicity of landscape forms connected with man's activities, forcing us to assign every affected landscape to the 'cultural' category, which is illogical. It would be more correct to divide all contemporary landscapes into a number of classes, depending on the degree and the character of changes introduced by man's activity. A number of varying proposals exist in this area (e.g. those of V. L. Kotyelnikov, D. V. Gogdanov, S. V. Kalyesnik, A. G. Isachenko and I. M. Zabyelin). By interpreting these proposals, it is possible to outline the following major classes of contemporary landscapes:*

1. *Unaltered or primitive landscapes* never or very rarely visited by man (e.g. Antarctic landscapes, many high-alpine landscapes, etc.).
2. *Slightly-altered landscapes* where man's activity has affected individual components (e.g. the animal world, through hunting and fishing) but where the principal natural relationships have not been disturbed. Among such landscapes are certain tundra, tayga and desert landscapes not yet included in man's agricultural activities.
3. *Disturbed (strongly altered) landscapes* subjected to prolonged but crude and irrational interference which has resulted in a substantial disruption of natural relationships and has altered the structure of

* It is Hartshorne's belief that more appropriate modifiers are 'wild' and 'tame' to distinguish 'the unaltered natural landscapes and those altered but uncontrolled by man, we may call "wild landscapes", in contrast to the "tamed" or "cultivated landscapes" of areas under man's control'. (Hartshorne, 1939, p. 348)

the landscape in a direction unfavourable to man. Landscapes in this group are particularly numerous. They are widespread in the tundra, where excessive reindeer grazing leads to the destruction of the lichen cover leaving large areas of barren land, and in the tayga where land clearing and timber felling are widespread. Similar activities are carried on in other zones as well.

4. *Transformed or cultural landscapes* in which the natural relationships have been completely altered by man's prior settlement and application of science, and by his rational distribution of cultivated lands, reconstruction of drainage systems and transformation of microclimates (e.g. with the aid of reclamation, forest planting, etc.) with the aim of securing the fullest and most effective utilization of natural resources, their preservation and development. Cultural landscapes of long-standing are distinguished by a high level of biological productivity, intensive biogenic cycles, favourable water budget, and complete exclusion or significant reduction of undesirable processes such as gully erosion, soil erosion, swamp development, etc. The morphological structure of a cultural landscape must also be transformed in the most useful direction through such redistribution of cultivated lands as would allow the most efficient utilization of individual sites (e.g. those suitable for orchards, grain cultivation, vegetables, etc.); in each case the mutual relationships between such sites should be considered.

It is natural to expect that socialism and communism will open the broadest possibilities for a planned transformation of landscapes on a scientific foundation. In the words of D. L. Armand:

> Under a planned socialist economy, the paths of natural processes progressively diverge from the natural and are transformed directionally. Simultaneously, however, if the programs implemented are well-founded in science, the boundaries of the transformed geocomplexes approximate the natural, since the transformation of nature and the differentiation of the economic profile take account of the natural potential of every locality.[89]

On the one hand, morphological contrasts may be smoothed out in some landscapes; on the other hand, structures of much greater complexity may be created (e.g. when farming lands, orchards, forest plantations, ponds, etc. are developed on a continuous upland area). Man's activities, therefore, erase some natural boundaries, create others, and obscure the distinctness of still others. It is necessary to add that the boundaries of transformed geocomplexes are, as a rule, much more distinct than the natural boundaries.

The study of so-called *city landscapes* constitutes a separate problem. The view has been expressed that landscapes disappear in closely built-up areas of large cities; it is difficult to agree with this view. Certainly landscapes are maximally transformed within the limits of large cities. Natural biocenoses are completely destroyed and are partially replaced with planted vegetation, often entirely untypical of the original landscape. The mesorelief is often entirely transformed in city areas owing to cutting and filling; the drainage pattern is also substantially altered. In Moscow, for example, only fourteen out of forty-five rivers and streams for which records exist now remain; sixteen were filled in and the remainder run in closed culverts or have been radically transformed. Permanent underground networks, sewage and water-mains sharply alter the groundwater systems. The soil is replaced with a so-called 'cultural horizon', including various street surfaces. The special character of the underlying surface, the predominance of stone, bitumen and concrete surfaces, as well as higher levels of atmospheric pollution, affect the climate. In particular, the air temperature in large cities is higher than over open terrain; microclimates are particularly strongly affected.

Nevertheless, 'nature' is not destroyed entirely in a city and the most important features of the natural landscape continue to act and exert their influence on the city and its inhabitants. It is difficult to imagine Leningrad, for example, without its characteristic climate, without the Neva and its numerous tributaries, without the flat, lowland topography, super-saturated mobile soil, the nearness of the sea, and many other factors wholly characteristic of this landscape. These factors determined in the past and continue to determine the specific features of building construction and municipal economy as well as the external appearance of the city.

V. V. Pokshishevski is fully justified in maintaining that all the natural components of a city (e.g. the climate, groundwater, etc.), even when powerfully transformed, do not cease to develop in accordance with physical-geographic law, and in advocating the *physical-geographic* investigation of cities as an indispensable element in urban development.[40]

In conclusion, it is necessary to stress that the study as well as the more detailed typology of landscapes altered by man's activities should be based on a natural classification of the original (reconstructed) landscapes. The character of any transformed or altered landscape is closely associated with the original natural landscape. The cultural landscapes of the tundra and the desert, the mountains and the plains, loessic rises and frontal apron lowlands will always differ from one another. To each type of natural landscape there corresponds a characteristic sequence or series of derived landscapes, the existence of which is due to man's interference. It follows that

measures aiming at rational utilization and transformation of terrain
should be implemented differentially, with regard to the various
types of natural landscapes. To do this, a classification of natural
(reconstructed) landscapes is needed.

The classification of landscapes

Landscape classification constitutes one of the most pressing and yet
increasingly complex problems in landscape science, the solutions to
which are only now being approached in connection with the compilation of survey (small-scale) landscape maps. Landscapes cannot
be classified 'from above' i.e. on purely theoretical grounds. A classification must be based on the differentiation, investigation and comparison of concrete landscapes and their subsequent arrangement into
groups, taking account of their common origin, components and
morphological structure. For this purpose we need to compare landscape maps, since cartographic processes themselves are based on
the maximum possible range of factual material and must assign each
landscape unambiguously to some specific slot in the taxonomic
schema. Maps allow neither gaps in scientific classification nor
ambiguous solutions but serve as the criteria of completeness and
objectivity in any geographic classification. As a result, in geography
the problem of classification is always directly bound up with mapping. The principles of landscape classification adduced below are
based on the experience gained in compiling a landscape map of the
U.S.S.R. The system proposed has a purely preliminary character and
requires considerable refinement. In addition, it should be borne in
mind that the classification was developed for maps with a specific
scale (1 : 4 000 000) and that in the future, when more detailed maps
are compiled for all the territories of the Soviet Union, the system will
have to be reconsidered and refined.

The classification of landscape has a multi-level character, i.e. it is
based on a taxonomic system of subordinate units. The higher-level
classificatory unit is landscape *type*, embracing landscapes which
share the most general features of genesis and structure and therefore
the fundamental geographic processes, in particular seasonal rhythm.
Since the character and intensity of the fundamental processes occurring in landscape depends primarily on climatic conditions, landscape
types must obviously have a specific zonal colouring. The character
of zonality, however, is not uniform in the various areas (sections)
of continents. It follows that landscapes of each type are distributed,
as a rule, within a single section of a continent. Some types are associated with oceanic sections (e.g. the western-European broad-leaved

forest type, the eastern-Asian subtropical type, etc.); others are associated with continental sections (e.g. the central-Asian desert type, the
eastern-Siberian tayga permafrost type); still others have a transitional
character (e.g. eastern-European landscape types—forest-steppe, subtayga or mixed forest). The assignment of a landscape to a specific
section is decided on the basis of certain genetic characteristics, the
development and structure of its base, and its previous history.

Every section of a continent, therefore, has a characteristic set of
landscape types. This does not mean that in certain sporadic cases,
depending on the particular combination of local conditions, certain
landscapes cannot appear beyond the boundaries of 'their' particular
section. Thus the landscape of the central-Kamchatkan depression
belongs to the continental type of eastern-Siberian tayga landscape,
even though it lies within the oceanic margin section of the continent.

A single type may include both lowlands and mountain landscape.
In mountain landscapes the general zonal-sectional conditions are
most strongly transformed by topography, yet they continue to assert
themselves and they determine the structure of vertical stratification.
This last feature enables us to assign a given mountain landscape to a
particular type. It must be admitted that the structure of the vertical
landscape stratification in the mountainous regions of the U.S.S.R.
has not been sufficiently studied, and this circumstance seriously
hampers the development of a general classificatory system of Russian
landscapes.[41] With regard to the plains, the predominant well-drained
types of vegetation and soils represent the most distinctive features
of each landscape type. It follows that landscapes of a single type are
characterized by common zonal-sectional structural features along the
plains and by the appropriate spectrum (type) of vertical stratification
in the mountains.

Over the territory of the U.S.S.R., landscapes of the following basic
types are found: arctic, European-Siberian tundra, far-eastern tundra,
eastern-European tayga, western-Siberian tayga, eastern-Siberian
permafrost tayga, far-eastern tayga, Pacific Ocean subarctic-subtayga
(Kamchatkan), eastern-European subtayga, western-Siberian subtayga, far-eastern subtayga, eastern-European broad-leaved forest,
far-eastern broad-leaved forest, eastern-European forest-steppe,
western-Siberian forest-steppe, eastern-European steppe, Kazakhstan-
Siberian steppe, Mongolian-Daurian steppe, Caspian-Kazakhstan
semidesert, central Asian desert, pre-Asian desert-steppe, mediterranean, Caucasian arid-forest and Girgan-Colchida wet forest. The
steppe, tayga and similar landscapes in various areas are regarded as
zonal analogues and are not included in a special taxonomic category.

Almost all landscape types are subdivided into *subtypes*; this

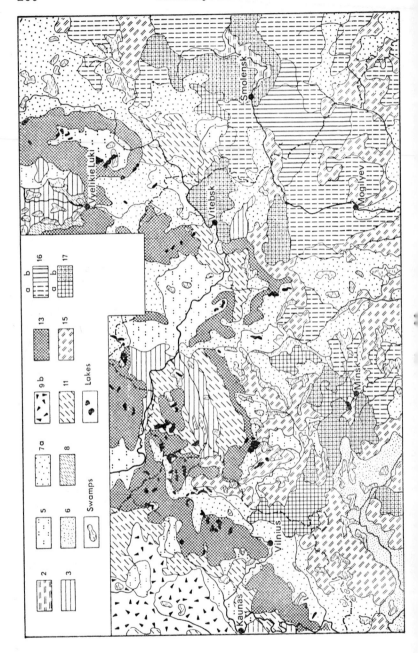

Fig. 36 Section of the landscape map of the U.S.S.R. showing eastern-European subtayga landscapes

Eastern-European subtayga (lowland) landscapes[42]

2, alluvial-deltaic low-lying plains, with floodplain meadows, floodplain marshes and black-alder stands; low level of utilization. *3,* low-lying flat, swampy, lacustrine-glacial plains, poorly exploited (usually less than one-fifth of the area under cultivation); clay and loam with swampy spruce forest and sphagnum moss. *5,* sands, superimposed over moraine, with swampy pine forest on peaty-podsolized-gleyey soils, together with red bilberry pine forest on poorly podsolized soils with sphagnum and grass swamps. *6,* low-lying flat, frontal, apron-alluvial plains, with thick sandy horizons, pine forest (in part broad-leaved pine) on soddy, weakly podsolized soils, grass and hypno-grass swamps, black-alder stands; elevated patches of sandy loam and rubble in places; in the past covered with broad-leaved and spruce forests; less than one-sixth to one-fifth under cultivation. *7a,* Superimposed over moraine (often at low depth), with large areas of swampy pine forest, sphagnum and sedge-grass swamps, with patches of dunes; varying levels of exploitation (in places less than one-tenth, elsewhere up to one-fourth and more under cultivation). *8,* low-lying, abraided poorly-drained morainic plains in the Valdai glaciation region, partly covered with a thin layer of lacustrine-glacial sands or sandy loams, with remnants of spruce forests and broad-leaved and spruce forests, together with secondary small-leaved forests; carbonate-less stony loam, with soddy, moderately and strongly podsolized and swampy-podsolized soils; poorly exploited (one-sixth to one-fifth, locally up to one-half, under cultivation). *9b,* a carbonate moraine, in part heavy lacustrine-glacial loams and clays, with soddy-carbonate podsolized, leached and soddy-gley soils; strongly exploited (more than one-half under cultivation). *11,* slightly elevated morainic plains in the Valdai glaciation region, with oak and spruce forests and broad-leaved and spruce forests (largely cleared) together with swampy spruce stands; rubble, partly weakly carbonate loam with podsolized, soddy-podsolized and swampy-podsolized soils; one-fourth to one-half under cultivation. *13,* monticulate-morainic uplands in the Valdai glaciation region, with remnants of kame, frontal apron, a multitude of lakes and swampy hollows, together with markedly higher rainfall, oak and spruce forests on mottley, predominantly strongly and moderately podsolized soils; varying level of exploitation (one-tenth to one-fourth, locally greater). *15,* slightly elevated rolling morainic plains in the Moscow glaciation region, largely leached and covered with a thin layer of sandy loam, with broad-leaved and spruce and secondary small-leaved forests on soddy, moderately and strongly podsolized soils; between one-third and one-half under cultivation. *16,* slightly elevated rolling and gently hilly plains in the Moscow glaciation (partly also in the Dnepr glaciation) region, with a cover of dusty loams, together with broad-leaved and spruce (largely cleared) forest on soddy moderately and strongly podsolized soils. *16a,* with thin (on moraine) carbonate-less or poorly carbonate loams; between one-third to one-half under cultivation. *16b,* with thick loessic poorly carbonate loams; one-half and more under cultivation. *17,* hilly elevated plains in the Moscow glaciation region, with erosion dissection and areas of morainic hills, secondary small-leaved (in place of broad-leaved and spruce) forests. On soddy moderately and strongly podsolized soils; varying level of exploitation (one-fifth to one-half). *17a,* rubble loams and sandy loams. *17b,* a cover of carbonate-less (partially loess-like, poorly carbonate) dusty loams over moraine.

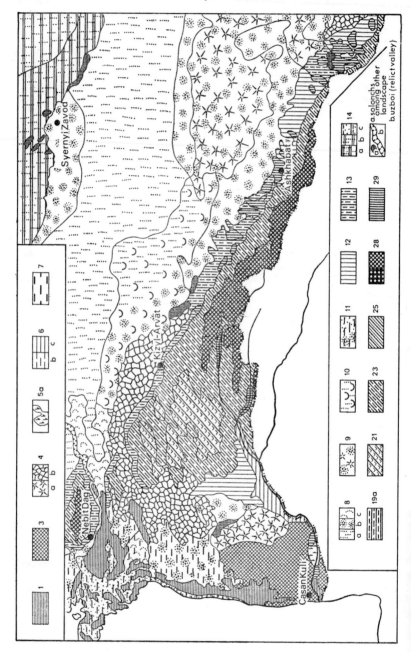

Fig. 37   Section of the landscape map of the U.S.S.R. showing the western part
of the central-Asian desert landscapes
Central-Asian desert landscapes (eastern desert)
Lowland landscapes: *1*, coastal solonchak low-lying plains, with succulent
thistles on solonchaks, and with saline sands with desert scrub and sparse
ephemerals. *3*, alluvial-deltaic low-lying solonchak plains, with Russian thistle
and other succulent-thistle and takyr-soil areas. *4*, low-lying takyr plains, with
blue-green algae and lichens on typical takyr soils, with local areas of wormwood and thistle-plant communities on takyr-like soils and solonchaks; *4a*, old-
alluvial; *4b*, proluvial. *5a*, turgays with tamarisk, elaeagnus and poplar cover,
with halophytic (*Aeluropus litoralis*) and coarse-green meadows on solonchaks,
meadow-solonchak and meadow-soils. *6*, old-alluvial and old-deltaic sandy-
clayey plains, with takyr-type soils; *6b*, with saltwort and southern wormwood
communities (*Salsola arbuscula, Artemisia herba-alba*); *6c*, with grey wormwood (*Artemisia terrae-albae*). *7*, old-alluvial slightly elevated plains, with
southern wormwood (*Artemisia turanica*) communities on dusty sandy loams.
8, old-alluvial and old-deltaic low-lying sandy wind-swept plains, with thick
carbonate non-saline sands and takyrs in depressions; *8a*, predominantly thick
sands with sandwort-sedge vegetation; *8b*, semi-fixed sands with psammophytic
scrub (saltwort *Salsola richteri, Calligonum* species, *Ammodendron* species,
etc.); *8c*, fragmented mobile sands, with a segmented cover of pioneer psammophytes (certain species of *Ammodendron, Astragalus, Calligonum, Eremospar-
ton*, etc.). *9*, old-deltaic low-lying sandy-takyr plains, with sandwort-sedge haloxylons on fixed sands, with algae and lichens on takyrs. *10*, old-deltaic low-
lying sandy-solonchak plains, with sandwort-sedge haloxylons on fixed sands,
with halochnemum and other succulent thistles on solonchaks. *11*, marine
accumulative low-lying sandy wind-swept plains, superimposed over saline
Caspian deposits, with psammophytic scrub (*Salsola richteri*, and species of
Astragalus and *Calligonum*) on semi-fixed sands. *12*, proluvial sloping sub-
montane loessic plains, with piedmont low-grass (semi-savanna), composed
mainly of meadow grass (*Poa bulbosa*) and sedge (*Carex pachystilis*), on light
serozems. *13*, proluvial sloping sub-montane rock-waste and shingle plains, with
ephemeral wormwood (*Artemisia sieberi*) on poorly developed rock-waste
serozems. *14*, stratified arid-denudation plains on Pliocene sandstones (in part
also clays and marls), with a cover of wind-swept sand; *14a*, elevated ridge
plateau with sandwort-sedge haloxylons on sand ridges, wormwoods (*Artemisia
terrae-albae* and *Artemisia herba-alba*) on thin layers of loose sand, wormwood-
thistle groups (*Salsola rigida*) on sandstone outcrops, with grey-brown solonets
rock-waste, and sometimes gypsiferous soils; *14b*, low-lying plains, with a
somewhat thicker sand cover, mainly with sandwort-sedge haloxylons and
some ephemers on loose sandy serozems; *14c*, elevated sandy plains, with
sedge meadow-grass (semi-savanna) and *Calligonum*.
Mountain landscapes: *19a*, low foothills, with sedge meadow-grass cover on
thin light sandy serozems. *21*, low foothills of 'badland' type on clay and marl
saline Paleogenic rocks, with a segmented cover of thistles and southern wormwood (*Artemisia herba-alba*) on serozems. *23*, low arid-denudation hills of
faulted Cretaceous and Tertiary rocks, with southern wormwoods on mountain
serozems. *25*, low and medium hills, with arid denudation and eroded cover,
with highland coarse-grass (semi-savanna) (including *Agropyrum trichophorum,
Hordeum bulbosum, Inula grandis, Prangus pabularia* and species of *Ferula*) on
mountain serozems, in combination with xerophytic, scrub, maple, juniper,
pistachio and almond woodland on cinnamonic and cinnamonic-brown soils;
25a, on sedimentary Paleozoic and massif-crystalline rocks; *25b*, mainly on
Tertiary and Mesozoic sedimentary rocks. *29*, oases, with irrigated soils on
alluvial-deltaic and proluvial foothill plains and mid-mountain depressions.

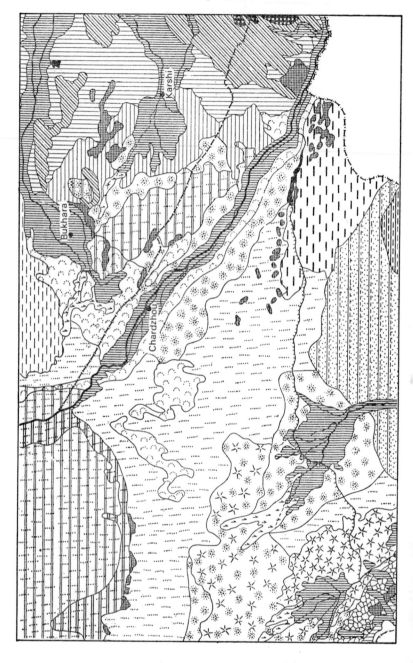

differentiation takes account of the secondary or transitional zonal factors in the structure. For example, three subtypes in the various types of tayga are distinguished: northern tayga, middle tayga and southern tayga. In the central-Asian deserts two subtypes are distinguished: northern and southern.

On the next level, the alpine and the lowland landscapes are divided and are regarded as individual *classes*. The distinctive features of mountain landscapes are the horizontal stages and the vertical stratification. Attention has already been drawn to the fact that despite its great diversity, the nature of a mountain landscape depends directly on the latitudinal and longitudinal position of the mountain range. The tundra and desert mountains or the mountains in the far-east of central Asia differ considerably with regard to their geographic conditions. By contrast, the mountain and lowland landscapes which belong to a single type (e.g. the central-Asian desert or eastern-Siberian permafrost-tayga) which are related genetically, are subject to certain general laws and are frequently associated by way of gradual transitions. For this reason alpine and lowland landscapes are regarded as second-order taxonomic categories, subordinate to the corresponding types and subtypes of landscape (e.g. the landscape of the Caucasus, which belongs to eight different types).

The next taxonomic level includes the *subclasses* of landscapes. Plain landscapes divide into two basic subclasses: low-lying and elevated. They differ not only with respect to their hypsometric position but also with respect to the degree of dissection, the genesis of topography and landscapes, the character of the substrata climate and other components. Low-lying landscapes are, as a rule, more recent, the relief has an accumulative origin and the bedrock is friable; elevated landscapes are usually older, denudation processes predominate and the bedrock often plays an important role in soil formation. Low-lying landscapes receive less rainfall, but they also are more poorly drained than elevated landscapes. All these factors are clearly reflected in soils and in biocenoses. In separating low-lying from elevated landscapes it is sometimes necessary to differentiate a transitional group of 'slightly-elevated' landscapes.

The division of alpine landscapes into subclasses reflects the stage differences in their characters. Three principal subclasses of alpine landscapes exist: low- , middle- , and high-alpine. The specific characteristics of each of these three subclasses have been discussed in an

Fig. 38 Section of the landscape map of the U.S.S.R. showing the eastern part of the central-Asian desert landscapes (see legend to figure 37)

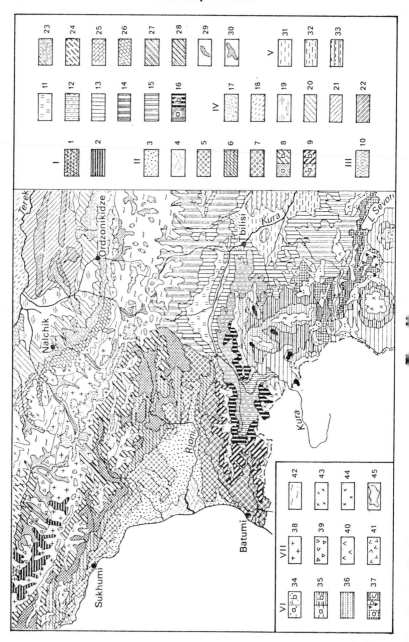

Fig. 39 Section of the landscape map of the U.S.S.R. showing the western part of the Caucasus

I, central-European forest barrier foothill landscapes (middle altitude)
1, fold-block ranges of Paleozoic sandstones, slates and limestones, with beech forest on mountain-forest brown soils. *2*, linear depressions in Jurassic sand-clay horizons and lateral valleys along high granitic ranges, with dark-conifer forests on mountain-forest brown podsolized soils.

II, Colchidian sub-tropical rain-forest barrier foothill landscapes
Low-lying plain intermontane landscapes: *3*, alluvial clayey plains, with Colchida broad-leaved forests (largely cleared) on podsolized yellow and alluvial soils. *4*, alluvial plains, with low-lying swamps and alder stands on peaty-gley, silt-marsh and meadow-boggy soils. *Piedmont-foothill landscapes*: *5*, rolling foothills of undulating Tertiary gravels, conglomerates, slates, limestone and tufogenic rocks, with Colchida broad-leaved forest (largely cleared) on yellow, red and partly soddy-carbonate soils. *Mid-mountain landscapes*: *6*, limestone ranges, with deep erosion dissection and karst forms, mainly beech forest on mountain soddy-carbonate soils. *7*, granitic areas (partly covered with poorly dislocated Tertiary sediments), with beech forest on mountain-forest brown soils. *8*, Jurassic sand-slate and volcanic ranges, with longitudinal depressions and deep lateral valleys; *8a*, predominantly beech forests on mountain-forest brown soils; *8b*, predominantly dark-conifer forests on mountain-forest brown podsolized soils. *9*, tufogenic and flysch Paleogenic rock ranges; *9a*, mainly beech forest on mountain-forest brown soils; *9b*, mainly dark-conifer forest on mountain-forest brown podsolized soils.

III, Transcaucasian arid-forest landscapes
Elevated-plain intermontane landscapes: *10*, alluvial-proluvial loamy and gravel plains, with low-land oak forests (largely cleared) on alluvial and meadow-forest soils. *11*, elevated synclinal depressions, with thick fluvio-glacial deposits partly covered with loessic alluvium and proluvium, with remnants of Georgian oak forests, thorny scrub and beard-grass steppe on cinnamonic and meadow-cinnamonic soils. *12*, synclinal elevated basins on Paleogenic volcanic and sedimentary rocks, with Georgian oak forests, secondary scrub and beard-grass steppe on cinnamonic soils. *Low-mountain landscapes*: *13*, strongly denuded foothills and low hills of Neogenic sedimentary deposits including conglomerates, Paleogenic flysch and Mesozoic volcanic rocks and limestones, with Georgian oak forests, secondary scrub and beard-grass steppe. *Mid-mountain landscapes*: *14*, folded flysch ranges, with deep lateral valleys, beech forests on mountain-forest brown (partly on mountain soddy-carbonate) soils. *15*, fold-fault ranges on Mesozoic volcanic rocks and limestones with granitic intrusions, beech forests on mountain-forest brown (partly on mountain soddy carbonate) soils. *16*, fold-fault ranges on Paleogenic flysch and volcanic deposits; *16a*, mainly beech forest on mountain-forest brown soils; *16b*, mainly pine or spruce forest on mountain-forest brown podsolized soils.

IV, eastern-European steppe landscapes
Lowland landscapes: *17*, alluvial clayey plains, with mottley-grass steppes on carbonate chernozem soils (cultivated). *18*, elevated plains, with loessic loams and dry grassland steppe on chestnut carbonate soils (a large proportion under cultivation). *19*, piedmont depressions on thick fluvio-glacial gravel, covered with loessic clays and loams, with meadow and meadow-steppe vegetation on alluvial and meadow-chernozem soils (a large proportion under cultivation). *20*, piedmont sloping terraced plains covered with loessic clays and loams, with piedmont meadow-steppe, steppe grassland and areas of oak (*Quercus robur*) forests on ordinary, leached, partly podsolized chernozems (a large proportion under cultivation). *21*, piedmont sloping plains and synclinal valleys with loessic

clays and loams, mottley-grass and grass steppes on carbonate chernozems (a large proportion under cultivation). *Low-mountain landscapes*: 22, frontal anticlinal ranges on Tertiary limestone-sandstone and clay deposits, with loessic loam and clay overburden, with mottley-grass and grassland steppe on carbonate chernozems (a large proportion under cultivation). 23, monoclinal-folded and anticlinal ranges on Tertiary conglomerate and sandstone, with longitudinal depressions in the sand-clay horizon on which oak groves (*Quercus robur*) predominate; mountain-forest brown podsolized soils. 24, longitudinal depressions in the Jurassic sand-clay horizon, with oak groves (*Quercus robur*) predominating on mountain-forest brown podsolized soils. *Mid-mountain landscapes*: 25, cuesta ridges on carbonate rocks, with steppe-type mountain grassland and areas of mountain steppe on mountain chernozems. 26, cuesta ridges on carbonate rocks, with longitudinal depressions in sand-clay deposits, mainly beech forest on mountain-forest brown and mountain soddy-carbonate soils. 27, monoclinal-folded limestone ridges, mainly with beech forest on mountain-forest brown soils. 28, fold-fault ridges on Paleozoic sedimentary rocks, with beech forest on mountain-forest brown soils. 29, mountain valleys (in the 'barrier shadow'), with dry steppe and mountain xerophyte complexes. 30, mountain valleys (upper zones), with pine forests.

V, eastern-Transcaucasian desert-steppe landscapes:
Intermontane plain landscapes: 31, piedmont alluvial plains, with beard- and feather-grass steppe on chestnut (partly solonets) soils. 32, sloping piedmont plains, with wormwood piedmont semidesert on grey-cinnamonic soils.
Low-mountain landscapes: 33, low hills on Neogenic sandstones, clays and conglomerates, with closed basins, mainly beard- and feather-grass steppe on chestnut (partially developed locally) and chernozem soils.

VI, pre-Asian desert-steppe landscapes
Intermontane plain landscapes: 34, alluvial plains; 34a, with mountain grasses and mottley-grass steppe on typical and carbonate mountain chernozems; 34b, with mountain grassland steppes on mountain leached-chernozems. *Mid-mountain landscapes*: 35, lava plateaux; 35a, with mountain grasses and mottley-grass steppe on typical and carbonate mountain chernozems; 35b, with mountain meadow-steppe on mountain leached-chernozems. 36, volcanic massifs and chains, with mountain-steppe meadows on mountain-meadow chernozem-like soils. 37, folded and fold-fault ridges on Paleogenic and Mesozoic volcanic, partially crystalline rocks and limestones; 37a, with mountain (mainly meadow) steppes on mountain chernozems; 37b, with juniper and xerophytic oak woodland on mountain cinnamonic soils; 37c, with mountain-steppe meadows on mountain-meadow chernozem-like soils.

VII, high-altitude mountain landscapes
38, high ranges, with acidic crystalline rocks and glacial and nival cover, with low-grass alpine meadows on mountain-meadow peaty soils, rocks, talus and glaciers. 39, high ranges with Paleozoic sandstones, slates and limestones, glacier and erosion cover, with subalpine and alpine meadows on typical mountain-meadow and peaty soils. 40, high ranges with Jurassic volcanic and Paleogenic tuff-flysch horizons, glacial and erosion cover, with subalpine and alpine meadows. 41, high limestone ranges, with subalpine and alpine meadows on mountain-meadow and chernozem-like soils. 42, high ranges and longitudinal depressions in Jurassic sandstones and slates with alpine and subalpine meadows, patches of wasteland, rocks, talus and glaciers. 43, high limestone cuestas, with subalpine and alpine meadows on mountain-meadow chernozem-like soils. 44, volcanic high ranges, with mountain meadows on mountain-meadow soils. 45, glaciers.

earlier section. In addition, we differentiate into individual subclasses the landscapes of mid-mountain depressions and of high-alpine plains (e.g. in Tibet and in eastern Pamir, etc.).

The lowest level in landscape classification is the *species*. These group together landscapes with a similar genesis, structure and morphology. Important features of landscapes belonging to a single species are uniform topography and substrata. Sometimes genetically similar landscapes exhibit secondary differences, often associated with certain peculiarities of the base (e.g. differences in the depth of the bedrock stratum or in the carbonate level). It is justifiable, in such cases, to distinguish *subspecies* of landscapes.

The system of taxonomic units outlined above is best illustrated by concrete examples. For this purpose sections of the landscape map of the Soviet Union and excerpts from the legend have been included. The first example (Fig. 36) relates to the landscapes of the eastern-European subtayga type; the second (Figs 37 and 38) to the landscapes of the central-Asian desert type (southern subtype). It should be borne in mind that the map sections shown do not include all the subdivisions of the legend but only those which are valid for the actual sections shown. The last example (Fig. 39) illustrates the alpine landscape in a region of the Great Caucasus. Since this example includes landscapes of very many types, the description of which by species would require much more space than is available, the legend includes only those subdivisions which are illustrated on the map.

In all examples the units designated with arabic numerals correspond to landscape species, except in the alpine areas where, owing to problems of scale and the need for optimum readability, certain departures have been allowed. (For this reason, in particular, the high-alpine landscapes are separated into a distinct group and are not distributed into corresponding types.) Lower case letters designate subspecies of landscapes and, for the alpine landscapes in the Great Caucasus, individual high-altitude strata.

4

Physical-Geographic (Landscape) Regionalization

Principles underlying physical-geographic regionalization

Regionalization is an independent form of classification of objects and phenomena which comprise regular terrain associations, the properties of which depend on their geographic position. Although different objects, natural as well as socio-economic, can be regionalized, we shall confine ourselves to regionalizing geocomplexes, i.e. to a physical-geographic or landscape regionalization.

Not every kind of physical-geographic classification of terrain may be called a regionalization; and certainly not every boundary shown on a map represents a region, i.e. a result of regionalization. Essentially, regionalization constitutes a classification of the earth's surface in which the isolated regions retain their territorial homogeneity and internal unity due to their common history of development, geographic position, common geographic processes and spatial association of individual constituents.

Regionalization takes account of the qualitative similarity of individual terrain sections, but does not elevate it to the role of a decisive criterion. Constituent units of a regional entity may be very different; what is important is that they must constitute a genetically homogeneous and territorially interrelated system. One example is the Russian plain—a large regional physical-geographic unit—an area with a broad spectrum of diverse landscapes, from tundra to semi-desert. Other examples are landscape zones which group together mountain or lowland landscapes.

The unity of every region has a complex character; it manifests itself in atmospheric processes, the structure of the drainage pattern and the migrations of plants and animals. But in the final analysis it is due to the effects of the two major geographic principles: the principles of zonality and azonality.

The degree of uniformity and the closeness of territorial bonds may

210

differ. These are obviously greater in adjacent sections of terrain than in sections lying at some distance from one another. This gives rise to the multitude of levels of regionalization, i.e. a consistent differentiation of regional complexes of different order inserted, as it were, one into the other. It is necessary not to lose sight of the fact that at any level of regionalization we always deal with individual territorial objects.* Every higher or lower regionalization unit constitutes a concrete and unique† piece of terrain to which a specific geographic designation may be assigned. In this, regionalization differs from classification in general (or typology), the units of which represent species or types of areas and do not distinguish territorial affinity.‡

In typological classification of geocomplexes, we rely primarily on their qualitative similarity, regardless of any territorial affinity that may exist among them. For this reason a type or species may include geocomplexes lying at some distance from one another, e.g. marshland urochishcha, steppe and low-mountain landscape, etc.

Certain authors designate the classification of geocomplexes as 'typological regionalization' and regionalization proper as 'regional regionalization'. Since, however, typology and regionalization are

---

* Although this point has been commented on in chapter 1, it seems appropriate here to provide a quotation from Bunge:
'The discussion is now developed sufficiently to turn to the concept of the "region as a concrete unit object". In order to produce areal classifications of identical sort no matter what differentiating characteristics are considered, it is necessary that there exist a perfect areal correlation between all phenomena of human significance. This condition is not met on the earth's surface. As an alternative attempt to preserve the region as a concrete unit object, it is possible, but absurd, to insist on some one arbitrary areal classification as sacred and immutable.' (Bunge, 1966a, p. 19)
This quotation reflects for the most part the Western geographers' attitude to the region as a concrete object.

† Bunge (1966a, pp. 9-13; 1966b, pp. 375-6) attacks the concept of uniqueness of location. In his opinion locations are no more or no less unique than stones, colours, snowflakes, fingerprints, etc. In another article Bunge (1966c) discusses the classification of locations.

‡ Grigg, quoting a 1937 article appearing in *Geography*, recognizes two categories of regional systems: generic and specific.
'Generic regions are ". . . those which fall into types and may therefore be said to be of a generic character, all the representatives of a particular type resembling each other in certain essential respects, according to the criteria selected—e.g. climate, character of vegetation, or human use."
'Such regions may occur in different parts of the world, but location is not a property used in their classification. A specific region is, however, a region ". . . whose character is determined not only by the intrinsic conditions of the area in question but by its location and geographic orientation."' (Grigg, 1965, p. 477)
Thus, when Isachenko speaks of 'species or genus categories' he is referring to a generic regional system. When 'territorial affinity' is considered, the regional system is specific. This chapter deals primarily with the delineation of specific regions.

mutually exclusive, the term 'typological regionalization' is internally contradictory. The term 'regionalization' is irrelevant if it is addressed not to individual pieces of terrain but to species or types. Regionalization should only be understood as a differentiation of individual units. There is no purpose, therefore, in introducing the expression 'regional regionalization', which is tautological. The concepts of 'regional' regionalization and 'typological' regionalization were criticized as early as 1955 during the first All-Union Congress on Landscape Science and have virtually gone out of use since then.

It is now possible to supplement our original definition with the statement that physical-geographic regionalization is concerned with the explicit statement of individual physical-geographic differences which developed as a result of the action on the earth's surface of the zonal and azonal factors of geographic differentiation.

Regionalization should be regarded simultaneously as a process of subdivision and unification. On the one hand, we are concerned with a successive subdivision of the earth's surface into territorial complexes progressively more simple and homogeneous in the character of internal relationships. On the other hand, physical-geographic regionalization progressively integrates the earth's landscapes into territorial units of an increasingly high order. These two aspects of regionalization are in no way mutually contradictory, but in fact supplement one another. Regionalization may be fully comprehended only if we approach it simultaneously from above and below.* It is difficult to implement a regionalization schema in practice if we do not follow both these approaches.† The view of certain specialists that regionalization should only be carried out from above must be treated as purely arbitrary, since it is not based on experience. From 'above' we can only achieve an approximate, preliminary schema. For a more detailed regionalization it is necessary to proceed from the low levels upwards. These problems will be discussed in detail in the section dealing with methods.

The results of regionalization and classification of geocomplexes

---

* Regionalization 'from above', the division of an area into units, is termed *analytical regionalization* and was applied by Herbertson (1905) in his division of the world into a number of different types of regions, climatic, physiographic, etc. This method is contrasted generally with regionalization 'from below', the grouping of geographic individuals into units, and is termed *synthetic regionalization*. (Unstead, 1916) Grigg (1965, p. 173) feels that synthetic regionalization is very similar to classification, while analytic regionalization corresponds to logical division.

† Hartshorne (1939, p. 468) also argues that 'we proceed in both directions, by division of larger units into smaller, and by building from smaller into larger; in every case, though the two methods partially check each other, in part we are forced to make compromises between them.'

jointly reflect with great clarity the present state of our knowledge of the earth's landscapes and the state of the theory of landscape science. Regionalization constitutes to a large degree the sum total of our knowledge at this stage of the development of landscape science. As it is implemented, regionalization reveals gaps both in data and theories, thereby stimulating the development of geographic science.

Regionalization has a very diverse and practical significance. Substantial evidence of this will be provided in the following sections. Here it is sufficient to note that regionalization enables us to fully catalogue natural conditions and resources, providing the most important natural-scientific foundation for planning their utilization.

The principal stages in the development of physical-geographic regionalization in pre-revolutionary Russia and in the U.S.S.R.

Long before the appearance of theoretical concepts in geography, practical necessity compelled man, even in very early times, to construct an abstract classification of the earth's surface in terms of units. This classification was based on the earth's most easily observed external characteristics. Even during the nineteenth century there was a widespread tendency to arrange the material in geographic descriptions and textbooks either according to political-administrative boundaries or in terms of river-basins or major relief forms.

The practice of regionalization began to assume a scientific character only after the requirements of a developing capitalist production coincided with a sufficiently high level of geographic knowledge. In different countries these two prerequisites appear at different times. In general, however, it can be concluded that regionalization became a scientific problem only in the nineteenth century, even though individual geographers had developed ideas on this subject earlier. In Russia both the natural and the economic conditions considerably assisted the development of the country's regionalization. The entire history of the physical-geographic regionalization of Russia can be divided into a number of periods.

The first period (the end of the eighteenth to the first half of the nineteenth century) was characterized by the appearance of rather imperfect empirical regionalization schemata. The need for Russia's regionalization was well understood by V. N. Tatischev, even during the 1730s. Only towards the end of the eighteenth century, however, did the prominent social thinkers begin to stress the significance of natural regionalization. In 1791 the leading revolutionary writer, A. N. Radishchev, put forward the idea that the administrative division of the country should take into account the different natural, historical and economic conditions, and he proposed that Siberia should be scientifically divided into 'naturally defined districts'.

It was common at that time to divide European Russia into three latitudinal zones: northern, central and southern. This schema was employed in geography textbooks by A. F. Bishing in 1766 and S. I. Pleshcheyev in 1787. In 1807 the schema was rendered a little more precise, partly for teaching purposes, by E. F. Zyablovski who differentiated four instead of three belts, their boundaries running along parallels of latitude.

In 1818, K. I. Arsenyev divided the territory of the Russian Empire into ten 'spaces' by grouping together certain provinces. The basis for this classification was, in the author's words, 'a striking similarity of certain provinces with respect to climate, soil quality, natural production and the industrial activities of their inhabitants.'

In 1834 there appeared in the *Agricultural Journal* a significant article attributed to E. F. Kankrin, which described eight latitudinal belts within the borders of European Russia. Two of these belts were the glacial and tundra belts. The remainder differed largely with respect to the level of agricultural specialization e.g. forestry and cattle breeding; primitive corn and barley farming; northern permanent grain farming, including rye and flax; wheat and fruits; maize and grapes; olive trees, silk and sugar cane. Only a rough idea of the boundaries of individual belts can be gained since they were not shown in the accompanying map. In 1842 the Ministry of Finance published 'an industrial map of European Russia' which included four 'regions': forest, industrial, chernozem and pastures.

The studies described above were relatively primitive in character owing to a lack of factual material and also to theoretical limitations. The factors commonly taken into account in terrain subdivision were gross differences in climate (no accurate data could be produced at the time owing to the undeveloped network of meteorological stations), the most important agricultural activities and, in part, the character of natural vegetation.

All the regionalization schemata during this first period are complex in character; as a rule they group physical-geographic and economic characteristics and are therefore clearly eclectic. Nevertheless one can easily distinguish studies which are predominantly economic (K. I. Arsenyev) or natural-geographic (most of the remainder) with the zonal tendency clearly manifested in the latter group.

A special mention must be made of a remarkable work, E. A. Eversman's *The Natural History of the Orenburg District* published in 1840, which includes a well-based subdivision of this territory into three natural zones corresponding to semidesert, steppe and the mountain-forest steppe of the southern Urals.

The second period (1850–90) was short but occupies an important position in the history of natural regionalization. It was a period of rapid development in the natural sciences including geography, and of great progress in the geographic investigation of Russia and in the growth of new geographic ideas. (These were discussed in detail in the first chapter of this book.) During this period regionalization was stimulated not only by scientific or social objectives but also by the need to systematize the mass of data on natural conditions in Russia. The differentiation of geography was directly reflected in regionalization—it developed mainly into a study of *partial* or *branch* systems.

In 1851, Trautfetter published the first botanical geographic regionalization of European Russia (with respect to the predominating tree genera); there then followed the similar systems of N. A. Beketov and F. P. Keppen in 1885. N. A. Severtsov in 1877, and M. A. Menzbir in 1882, were the first to experiment with zoogeographic regionalization. In 1871, A. I. Voyeykov produced a climatic regionalization of the Great Caucasus. In 1886, S. N. Nikitin proposed the first geomorphological regionalization of European Russia, which was based on a division of the country into two major areas; glacial and non-glacial.

The period between 1850 and 1890 was thus characterized by the transition from very broad and complex regionalization schemata to a more detailed and more systematic classification on the basis of the study of individual natural components. This period of analytical works produced the necessary prerequisites for the subsequent transition to a synthesis on a higher level, which occurred during the next stage in the history of regionalization.

It should be noted, however, that the botanical-geographic and zoogeographic regionalizations of that time were not necessarily as narrow as might be supposed. Most authors took into account a variety of geographic factors including historical factors. Multi-level systems of taxonomic units were employed for the first time in Trautfetter's work—zones and districts—and in Keppen's work—regions and zones. Finally, the continued development during this period of the zonal principle of regionalization must be noted.

The third period (the end of the nineteenth and the beginning of the twentieth century) is associated with the work of V. V. Dokuchayev and his school. The development of regionalization during that time was based essentially on the zonal concept. The first maps of Russian vegetation and European-Russian soils to reflect the principles of zonality appeared at the turn of the century.

An important place in the history of regionalization belongs to G. I. Tanfilyov who published a physical-geographic regionalization of European Russia in 1897. Tanfilyov employed a number of variables, e.g. soils, bedrock and vegetation, as well as the character of the groundwater, the distribution of swamps and, to a significant degree, the history of landscapes. Tanfilyov's regionalization is based on a triple-level system of units. The highest level is the region (four are differentiated: the northern region, i.e. non-chernozem; the southern region or old steppe; the Aral-Caspian region and the southern coastal-Crimean region). Regions divide into zones and zones into districts.

This regionalization has substantial defects; in particular it fails to take into account hypsometric and geomorphological differences and it does not adequately justify the criteria used in the differentiation of regions; the regions themselves, as well as some of the zones, are frequently represented by broken boundaries. Today it is of interest largely as a historical document; in its time, however, it represented a considerable improvement over earlier systems.

The next attempt at the regionalization of European Russia was carried out by P. I. Brounov who in 1904 took relief as his basis. However, he failed to apply it consistently. The actual boundaries of provinces and regions were often drawn along the isolines of various climatic indices.

In 1907 A. A. Kruber proposed a regionalization system for the same territory, in which, like Tanfilyov, he differentiated regions, zones and districts. The merit of Kruber's system is that it reflects very clearly the peculiarities of topography and geological structure. His classification into zones, however, is more schematic than in Tanfilyov's system, and his differentiation of districts is perhaps ill-founded.

During this period, problems of agriculture and forestry generated much interesting research into the theory and methods of regionalization. To this work belongs the zonal division of the Russian plain which was developed in 1899 by G. N. Vysotski in connection with the afforestation of the steppe. Taking into account the mutual interrelation between climate, bedrock, soils and vegetation, Vysotski divided the Russian plain into four zones (excluding the tundra) and showed that each of these is characterized by a specific complex of conditions for forest growth; each, therefore, required an individual approach to the problem of afforestation. A little later Vysotski substantiated his system with his own objective index of moisture conditions (i.e. a ratio of rainfall to evaporativity). Despite its excessively general character and lack of precise boundaries, Vysotski's zonal system was the most thoroughly developed regionalization during the third period of the physical-geographic regionalization of Russia.

Vysotski was responsible for the idea of regionalizing from below, i.e. by the gradual amalgamation of the elementary natural units of terrain, and also for a number of other valuable ideas on the theory of regionalization.* In particular, he came out decisively against attempts to unify natural and economic regionalization within a single framework.

In 1908 V. V. Viner developed a system of natural zones for the whole of Russia as a basis for the planning of a network of experimental agricultural stations. He used as a basis the distribution of natural vegetation, relying on the hypothesis that this distribution reflects most accurately the effect of external conditions and provides the most complete available synthesis of all these conditions.

A whole series of works concerning regionalization and associated with the problems of afforestation were published during this period. The leading ideas in this area were those of G. F. Morozov who gave considerable weight to the zonality principle. In 1916 A. A. Kryudener experimented with the natural classification of European Russia for afforestation purposes. His system was based on six zones defined by the ratio of rainfall to evaporativity.

The first regionalization of the entire country was produced in 1913 by L. S. Berg. Berg designated his regionalization with the adjective 'landscape'. It has already been mentioned in chapter 1 of this book that it was Berg who introduced into geography not only the system of landscape zones but also the concept of landscape itself: he regarded landscape zones as groups of landscapes.

Berg's first system of Russian landscape zones included the following (1) the tundra zone; (2) the tayga plains (with the subzones conifer-tayga and tayga, with a certain proportion of broad-leaved species); (3) the forest-steppe (with the subzones forest-steppe on grey forest soils, and forest-steppe on chernozems); (4) the chernozem steppe (with the subzones lowland chernozem steppe and high-altitude Transbaykal chernozem steppe); (5) dry chestnut steppe (with the subzones dry-steppe plains and hilly dry steppe); (6) semidesert; (7) desert; (8) Amur lowland and Ussuri-land zone, with Manchurian-type forests. Berg treated separately the mountain regions with well-developed stratification.

During the first decade of the twentieth century there also appeared

---

* See Unstead (1916, 1926, 1932 and 1933).

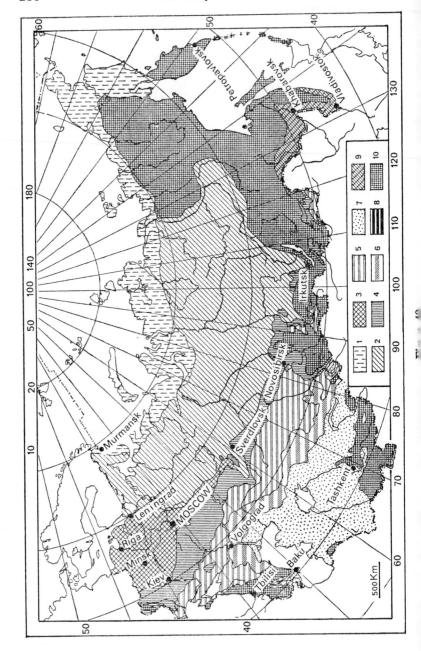

individual examples of a detailed natural subdivision of certain territories, in most cases provinces. The most interesting of these is the
study by S. S. Neustruyev, L. I. Prasolov and A. I. Byezonov, *The
Natural Regions of the Samara Province*, published in 1910. This
system was based simultaneously on the zonal principle and on the
analysis of the topography and bedrock, from the viewpoint of geological history.

Thus the period under discussion was not characterized by a return
to a synthesis in natural regionalization, but for the first time regionalization was based on theoretical principles, in particular on the
principle of zonality. In the area of branch regionalization (mainly
soil, geobotanical and, in part, climatic) there was a tendency to
augmentation and increased complexity, except for geomorphological
regionalization which developed independently.

The fourth period coincides with Soviet control. In its early stages
(during the 1920s) there was a growing interest in natural regionalization, due to the reconstruction of the political-administrative subdivision of the country, the establishment of Union Republics and
Autonomous Nationalities and the creation of economic regions under
the Gosplan (National Plan). This situation also gave rise to the
development of local regionalization systems embracing individual
republics (e.g. the Ukrainian S.S.R., the central-Asian Republics, and
Dagestan), provinces (e.g. Orenburg, Omsk, Voronezh and Odessa),
or large regions and areas (e.g. south-eastern European Russia, the
northern Caucasus, the Yenisey region, the Ural region, etc.).

Apart from a few exceptions, these extensive endeavours were on
the whole at a low theoretical level. Whereas it was entirely reasonable
to restrict the general systems of the natural subdivision of the whole
of Russia, or even of its European part, to the differentiation of zones
and subzones and of the principal mountain regions, such a practice
proved to be completely inadequate for the detailed regionalization of
small territories. The principles and the methods of detailed intrazonal regionalization were, however, not as yet developed. A
theoretical foundation for such a regionalization is provided by landscape science which, during the 1920s, was still in its initial stages.

The most important contribution to the theory of regionalization
was the derivation of the *principle of 'provinciality'** which very

* See chapter 2, p. 69.

Fig. 40 Landscape zones of the U.S.S.R. (after L. S. Berg)
1, tundra; *2*, tayga; *3*, mixed forest; *4*, forest steppe; *5*, steppe; *6*, semidesert;
7, desert; *8*, subtropics; *9*, far-eastern broad-leaved and mixed forests; *10*,
alpine landscapes.

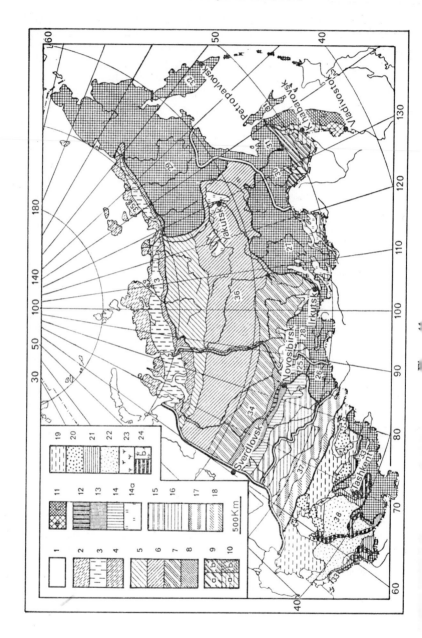

quickly gained widespread acceptance but was not immediately implemented.

During the thirties and the forties further regionalization systems were published embracing the entire territory of the U.S.S.R. as well as some large sub-areas; these were associated, in the main, with the publication of university textbooks on the physical geography of the U.S.S.R. Among these textbooks was Volume 1 of L. S. Berg's *Landscape-Geographic Zones of the U.S.S.R.*, published in 1931, which included a more refined system of landscape zones (Fig. 40) and their detailed description. Berg, however, did not develop an intra-zonal sub-division, and the principle of provinciality is not reflected in his work; for example, mountain regions are discussed, as before, separately from zones.

A still more detailed physical-geographic regionalization of the European part of the U.S.S.R. was published ten years later in 1941 by B. F. Dobrynin, but it cannot be regarded as a successful effort. On

---

Fig. 41   Landscape-geographic regionalization of the Asian part of the U.S.S.R. (after S. P. Suslov)

Zones and subzones: *1*, glacial zone; *2*, arctic tundra; *3*, typical tundra; *4*, forest tundra; *5*, northern subzone of western- and eastern-Siberian sparse forest; *6*, cedar-swamp western-Siberian tayga subzone and the central subzone of eastern-Siberian tayga; *7*, coniferous-swamp subzone of western Siberian tayga and the southern subzone of eastern-Siberian tayga; *8*, transitional zone of secondary aspen-birch swampy forest; *9*, transitional north-western region of the Amur-Maritime district with predominant (*a*) plain and (*b*) alpine deciduous forest, foggy (shrubby sphagnum bog) and individual areas of pine forest; *10*, subzone of broad-leaved plain (*a*) and alpine (*b*) Amur and northern Ussuri forests; *11*, southern Ussuri forests (*a*) plain and (*b*) alpine; *12*, floodplain-meadow-solonchak forest-steppe; *13*, northern forest-steppe; *14*, southern or typical forest-steppe; *14a*, forest-steppe islands in eastern Siberia; *15*, feather-grass and mottley-grass steppe on chernozems; *16*, feather-grass fescue steppe on chestnut soils; *17*, white-wormwood fescue semidesert on light-chestnut soils; *18*, fescue-wormwood semidesert on brown soils; *19*, grey-wormwood saltwort desert on clayey and stony brown-soils; *20*, northern-type sandy desert; *21*, occasional desert on loess; *22*, southern-type sandy desert; *23*, solonchak desert (large islands); *24*, valley landscapes (*a*) oasis and meadow and (*b*) large turgay patches. *Mountain regions*: *25*, Kuznetskiy Alatau, Salir and Kuznetsk basin; *26*, Altai; *27*, Transbaykal, Baykal and adjacent northern and north-eastern areas; *28*, Sayana and near-Sayana; *29*, alpine region of north-eastern Siberia; *30*, Amur-Maritime region; *31*, lower Amur-Okhotsk region and Sakhalin; *32*, Kamchatkan-Kurilian volcanic region; *33*, central-Asian alpine region; *34*, western-Siberian lowlands; *35*, western-Siberian arctic region; *36*, central-Siberian region; *37*, central-Asian semidesert (central Kazakhstan) region; *38*, central-Asian desert region.

District boundaries (e.g. western Siberia, eastern Siberia, far east, central Asia) are indicated by heavy lines; boundaries of regions by light lines.

the other hand, a regionalization of the Asian part of the U.S.S.R. published by S. P. Suslov in 1947 (Fig. 41) as part of his university course in the physical geography of the U.S.S.R., represents a notable contribution. Suslov's system reflects both zonal and azonal factors, and this distinguishes it from all the other similar schemata.

In 1938 the People's Commissariat of Agriculture of the U.S.S.R. approached the U.S.S.R. Academy of Sciences with a request for the rapid preparation of a natural-historical regionalization of the U.S.S.R. In his letter the Commissar, I. A. Benediktov, wrote to the president of the academy, V. L. Komarov, that 'we do not possess, as yet, a comprehensive, scientifically based, natural-historical regionalization of the U.S.S.R. which takes account of the major indices of the natural productive forces in agriculture: the climate, soils and vegetation'. Such a regionalization was urgently required, since without it one could not 'correctly develop or undertake a vast range of agricultural-industrial and administrative measures aimed at continuing growth of the State and Co-operative farm production, such as the introduction of appropriate rotational systems, soil-cultivation methods, fertilization, regional distribution of crops'.[1]

The Academy of Sciences established a committee whose task, the natural-historical regionalization of the U.S.S.R., was completed rapidly by several institutes of the academy. In addition, certain branching types of regionalization (e.g. geomorphological, hydrological and geobotanical), as well as complex ('natural-historical') regionalization systems were also developed.[2] Geographers took no part in these latter activities. One of the many positive aspects of the natural-historical regionalization of the U.S.S.R. was the attempt to combine zonal factors with azonal; however, the attempt was on the whole unsuccessful.

The contemporary stage in the development of the physical-geographic regionalization of the U.S.S.R. comprises the last ten to twelve years. The rising interest among geographers in theoretical investigations has also led to regionalization, in which different aspects of regionalization have been studied. The many special theoretical studies which have been published, as well as the publication of many textbooks on the physical geography of individual republics, provinces and large economic regions, prompted many regionalization schemata for individual territories. Furthermore, numerous large expeditions have carried out regionalization studies to satisfy practical needs arising from the cultivation of virgin lands, afforestation, large-scale hydro-electric power engineering, and other types of construction. These studies were concerned purely with complex regionalization, i.e. physical-geographic regionalization as such, and not with partial or branch regionalization, whose concern is individual components.

In 1956 universities undertook the huge task of the physical-geographic regionalization of the U.S.S.R. in order to provide a basis for the future planning of Soviet agriculture. A number of inter-university conferences were convened which evaluated results and adopted a system of taxonomic units for regionalization as well as certain general methodological principles. A joint effort of many groups of geographers is contributing to the emergence of uniform principles and methods of regionalization despite the fact that theoretical differences have not all been overcome and the individual schemata proposed for certain territories do not coincide.

The fundamental principles
of physical-geographic regionalization

The very large number of problems connected with physical-geographic regionalization may be divided into two groups. The first includes the *theory* of regionalization, i.e. the set of principles underlying the practice, and the second the *method* of regionalization, i.e. the implementation in practice of a regional subdivision of a given territory.

The most important methodological principle in regionalization is the recognition of its objective character. A system of regionalization units reflects the objective rules of physical-geographic differentiation, and is therefore independent of the aims and objectives which regionalization may be designed to satisfy.* These aims and objectives may be very different, yet these differences do not cause changes in the boundaries of natural regions. The intended purpose of a regionalization (e.g. agricultural, land-reclamation, educational, etc.) is only reflected in the degree of detail included (i.e. in the choice of the minimum taxonomic level), in the amount of description of delineated regions (i.e. the selection of appropriate indices of natural conditions) and, if necessary, in the regrouping of regions in pursuit of solutions to specific practical problems. If, for example, two or more regions require identical anti-erosion measures, they may be grouped together.

Thus every applied regionalization is an interpretation of one uni-

---

* The idea that physical-geographic regionalization is independent of aims and objectives is tied to the concept of the existence of regions. Grigg (1965, p. 482) takes the view that if the 'purpose of a system is of paramount importance in construction (of regional systems), many disputes about the respective merits of different systems could be resolved. Arguments about whether one system is better than another often become meaningless when we realize that the respective classifications have different purposes.'

versal physical-geographic regionalization.* Each such interpretation, as shown by S. V. Kalyesnik, is unconcerned with the principles of regionalization and merely utilizes its data for applied purposes. It follows that the theoretical basis of regionalization is the study of the laws of territorial physical-geographic differentiation. These laws underlie the entire system of taxonomic units of regionalization.

It has already been established in chapter 2 that the existence in nature of natural regional complexes of different orders is determined by the action of zonal and azonal factors; contemporary regional differentiation is a result of complex, time-mediated relationships between these and other factors. Every region has its history, and its unity springs from a common historical development in given zonal and azonal conditions. This in turn means that regionalization must rest on a *genetic* or *historical* principle.

Many geographers have stressed the importance of the genetic principle, but its practical application has been lagging. In part this is due to the incompleteness of geographic records and the difficulties associated with the reconstruction of the origins and the history of landscapes, districts, zones and other regional complexes. Furthermore, the theory of the genetic principle is as yet insufficiently advanced. The principle itself is often too narrowly conceived and is limited in application to the origin and the history of the geological base rather than to that of the geocomplexes themselves. Some geographers who justly criticize this trend move to the other extreme and completely reject the possibility of regionalization on a genetic basis.

To apply the genetic principle in regionalization is to explain the progress of physical-geographic differentiation of a territory, to show when and why the regional units of different orders have become differentiated, and to account for their common genetic origin. It has been noted earlier (chapter 2) how important the genetic approach is in the study of landscape zones. It is no less important in revealing the regional complexes of any other rank, and makes it possible to establish between them taxonomic relationships, i.e. a natural hierarchy.

* The existence of one universal physical-geographic regionalization is refuted by Hartshorne:
'The question as to which criteria shall be chosen for determining regions likewise finds no answer in nature, says Hettner. The choice must be made by "the geographer, according to his subjective judgment of their importance. Consequently one cannot speak of true and false regional divisions, but only of purposeful or non-purposeful. There is no universally valid division which does justice to all phenomena; one can only endeavour to secure a division with the greatest possible advantages and the least possible disadvantages".' (Hartshorne, 1939, p. 466)

Thus, following the establishment of the folded base, the Russian plain has, as a unique physical-geographic district, maintained a characteristic unity and a territorial homogeneity over many periods of geological history despite the fact that its constituent landscapes were continuously changing while its own area expanded and contracted. As a result, however, of differentiated epeirogenic movement, marine transgressions, non-uniform rainfall in different areas, continental glaciation and other processes, individual landscape or physical-geographic regions have become established, each of which comprises a complex of landscapes of similar origin and age. Most regions were already outlined at the beginning of the Quaternary, but their ultimate boundaries were defined much later. Examples are provided by the Polesye, central-Russian, Black Sea, and other regions. The north-western region, which coincides with the area of the most recent glaciation thereby reflecting its character, is the youngest region.

Parallel with the differentiation of regions which are essentially azonal formations was a simultaneous differentiation of the Russian plain into landscape zones and subzones. These were established within their present-day borders very much later than the landscape regions. The boundaries of the zones and subzones of the Russian plain are subject to much greater fluctuation than those of the regions.*

Finally, landscapes were formed within the framework of regions and zones (or subzones) at the most recent stage in the history of the Russian plain. In the north-western region, for example, this stage was characterized by glacial and fluvio-glacial accumulation, erosional and accumulative-depositional processes in drainage basins, frequent macroclimatic fluctuations and the introduction of Siberian and western-European flora and fauna. The interaction of these processes in different areas of the region varied considerably, leading to the development of different landscapes (concrete examples have been given in the preceding section).

Experience with the application of the genetic principle to regionalization has led to the following conclusions:

1.   In the course of the history of the geographic envelope its regional differentiation progressively grew in complexity. The larger territorial units were differentiated earlier than the smaller units; in other words, the lower units became established within the larger ones, and therefore must be more recent. Nevertheless, regions of the same order may be of different age. The north-western region of the Russian plain

---

* All these physical-geographic units are considered in the individual (regional) sense rather than the typological sense. See conclusion 3 below.

is younger, for example, than the central (Upper Volga) region, while the latter is younger than the Urals (High Transvolga).

2. Physical-geographic differentiation proceeds simultaneously along two channels: regional complexes are formed by the action of zonal factors on the one hand and azonal on the other. Both processes continuously interact, yet are entirely independent of one another since they arise from different causes.

3. Two types of genetic relationships and, therefore, two categorical steps of regional physical-geographic units can be proposed: zonal (zones and subzones) and azonal (districts and regions).

It must be emphasized that the units of both these categories are geocomplexes. Moreover, although the structure of any one of these units results from the operation of both zonal and azonal factors, the differentiation of regional complexes and the formation of boundaries are determined overwhelmingly by either the zonal or azonal group of factors. Each unit manifests a genetic unity of certain landscapes, yet the unity of zones and subzones on the one hand, and districts and regions on the other, arises from different causes and from diverse interrelationships.

The azonal interrelationships between landscapes are often much older than the zonal, since the principal element of these inter-relationships is the most conservative component: the solid base which determines the continuity between primitive and contemporary landscapes. Every azonal unit is characterized primarily by common tectogenesis and macrorelief; it serves as a stage on which certain contemporary geographic, broadly-acting processes, especially atmospheric processes, are realized. The terrain of each district or region creates characteristic conditions for atmospheric circulation and often serves as a focus for the formation or transformation of air masses, in some cases promoting the development of highly specific centres of atmospheric activity. Each region or district is associated with a characteristic set of landscapes (in particular, many districts, as will be shown later, are distinguished by characteristic spectra of latitudinal zones). The same zones or subzones are, as it were, differently articulated when they enter different districts or regions.

The unity of a landscape zone or subzone resides in genetic inter-relationships and processes of a different kind. The common character of the conditions for landscape development within a single zone are determined primarily by the radiation conditions and by supplies of moisture. Accordingly zonal unity is determined overwhelmingly by contemporary processes, e.g. climatic, geochemical, biogeographic, etc. Naturally landscapes belonging to a given zone or subzone exhibit a significant degree of structural homogeneity; for this reason landscapes of one type, or a small number of types constituting zonal similarities, predominate in every zone. Zonal units, by contrast with azonal, are more dynamically transformed. This accounts for the uneven character of spatial boundaries; physical-geographic districts and regions have more distinct and 'stable' boundaries than zones and subzones.

4. It follows that zonal and azonal classifications do not conform but, as it were, intersect one another and do not constitute a simple hierarchy. A 'focal point' at which they converge is the landscape, which constitutes the basic (smallest) unit of physical-geographic regionalization. Landscapes are simultaneously units of regions and also of zones (or subzones). All higher units of regionalization constitute nothing other than zonal and azonal groupings of landscapes. This accounts for the important role of landscape science as one of the two (alongside the study of the laws of territorial differentiation) most important theoretical bases of physical-geographic regionalization. The characteristic features of landscapes and, in particular, those which raise it to the level of a territorial unit were discussed in detail in an earlier chapter.

A landscape constitutes an important qualitative stage in the system of geocomplexes. From the viewpoint of the aims of regionalization, its essential property is its indivisibility with respect to either zonal or azonal characteristics. The concept of a landscape as the basic regional unit is important not only for the theory of regionalization but also for the co-ordination of all the different types of partial (branching) natural regionalization. Landscapes correspond in area to the smallest units in soil, geobotanical, hydrological, etc. regionalization systems, although their boundaries may not exactly coincide owing to the varying nature of spatial transitions in different components. Such correspondence is rarely observed on higher levels (the reasons for this are discussed below). Accordingly, as noted by V. B. Sochava, the acceptance of landscape as a basic unit in the natural classification of the earth's surface, necessitates the development of a unified system of genetic natural regionalization.

Neither facies nor urochishcha, strictly speaking, are regionalization units, since as has already been shown they do not always preserve the typical characteristics of all the higher regional complexes, nor do they provide information about the total set of natural conditions in a territory. Every region necessarily constitutes a regular and linked system of different territorial units. Not all geographers, however, share this view. D. L. Armand, N. I. Mikhailov and N. A. Solntsev maintain that all geocomplexes, including the facia, constitute categories in regionalization. They occasionally refer to the morphological classification of a landscape as intra-landscape or large-scale regionalization. If urochishcha are conditionally regarded as micro-regions, then the facies can no longer be included among regional categories, since they do not constitute a territorial combination of various geographic units.

Very often a *rayon* is accepted as the lowest-level unit of regionalization. However, virtually every geographer defines it in his own way; most often it is simply the last stage of regionalization which the author has reached in classifying or describing a territory. Therefore, in comparing regions derived by individual geographers, there can be found incredible variety and inconsistency. Precisely because of this indeterminancy, the term 'rayon' is very attractive to some workers. It does not restrict them to narrow taxonomic categories and enables them to differentiate regions purely on empirical evidence, i.e. on the existing factual material and the author's individual views on physical-geographic differentiation.

If the term 'rayon' is given a more precise definition, it is found to coincide with landscape, at least as far as the plains are concerned. In mountainous regions, on the other hand, the regional differentiation of which is much more complex and requires a larger number of taxonomic levels, one would like to differentiate between these two taxonomic units. However, it will be more convenient to return to this question later after studying the general system of taxonomic units and their application to the regionalization of various districts.

One of the central problems in the theory of regionalization is that of the system of taxonomic units. The existence of two types of regional units does not necessarily mean that geographers are free to choose either one of the two systems (zonal or azonal) and design regionalization schemata accordingly. It is generally accepted today that unit systems in regionalization should account for both these aspects simultaneously. Purely zonal schemata belong to history. An example of regionalization based solely on the azonal principle is the schema dividing all landforms into 'districts' proposed by A. I. Yanputnin.[3] N. A. Solntsev's regionalization of the European part of the U.S.S.R. has in fact been based exclusively on the azonal principle.[4] Finally, in the opinion of Yu. K. Yefremov, the zonal and azonal systems should have an independent status. (Yefremov defines regionalization proper as azonal classification, whereas the zonal system is regarded as 'typological'.)[5]

There is little doubt, however, that regionalization based on azonal characteristics is as one-sided as regionalization based on purely zonal characteristics, since both reflect only one aspect of the geographic differentiation process. In such one-sided systems the position of a landscape is as undetermined as the position of a point on the earth's surface when only one geographic co-ordinate is known.

The task, therefore, is to find ways of designing taxonomic systems which can combine zonal and azonal units. The difficulty resides in the fact that zonal and azonal classifications have independent origins and are not members of a single hierarchy.

It would not be true to say that there is absolutely no mutual interdependence between zonal and azonal differentiation. A partial interdependence certainly does exist; one example is the relationship between the distribution of zones and subzones and their boundary configurations and orographic and lithological factors, another is the partial coincidence of zonal and azonal boundaries, etc. These facts certainly do not conflict with the independent status of zonal and azonal regionalization units.

A large number of taxonomic systems combining both approaches has been proposed, and they can be broadly grouped into two diametrically different systems: *uniserial* and *biserial*.

The discussion of these systems must be postponed, however, until a better idea of the primary higher-level (i.e. above the level of landscape) categories of zonal and azonal classification, independently of taxonomic interrelationships, is gained.

Higher-order taxonomic units in physical-geographic regionalization

Some geographers, A. A. Grigoryev, V. B. Sochava, I. P. Gerasimov and others, regard the *belt*, which is the largest latitudinal-zonal element of the earth's surface classification, as the highest unit in regionalization. Many specialists consider, however, that there are no sufficient grounds for the differentiation of belts as complex physical-geographic units. There are no belts in *The Natural-Historical Regionalization of the U.S.S.R.* produced by the U.S.S.R. Academy of Sciences; they were not accepted by L. S. Berg, S. P. Suslov and other geographers. Belts are also excluded from a system of regionalization units accepted by the most recent inter-university conferences.

The criteria for the introduction of a belt as a category in physical-geographic regionalization have not yet been formulated. It is highly significant that among the different systems dividing continents or the territory of the U.S.S.R. into belts, it is difficult to find a pair the members of which would resemble one another. Thus A. A. Grigoryev divides the northern hemisphere into six physical-geographic belts: arctic, subarctic, temperate, subtropical, tropical and equatorial (the U.S.S.R. lies within the first four). According to I. P. Gerasimov, there are only four belts in the northern hemisphere: northern polar, boreal (temperate), northern-tropical-xerothermal (subtropical) and tropical hydrothermal (the territory of the U.S.S.R. lies within the first three belts). A similar system for the classification of the territory of the U.S.S.R. was proposed by V. B. Sochava. One of Sochava's most recent monographs, however, assigns the entire territory of the U.S.S.R., including the Arctic, to a single ('extra-tropical') belt.[6] The most recent soil-geographic regionalization of the U.S.S.R. produced by a group of soil scientists at the Institute of Soils in the U.S.S.R. Academy of Sciences, and the Council of Investigation of Productive Resources, divides the territory of the U S.S.R. into four belts.[7]

These examples indicate that belt-differentiation procedures are entirely unsatisfactory. The consistent pattern of change in zonal phenomena creates considerable difficulties for procedures grouping zones into larger territorial units, and such procedures eventually become redundant. Specifically, which zonal boundaries should be elevated to the role of especially pertinent (belt) boundaries? It is entirely natural that most belt systems are regarded as artificial. Yu. P. Parmuzin, for example, assigns the most northern part of the northern-Siberian tayga to the cold belt and the remainder to the temperate-cold belt. The authors of the soil-geographic regionalization of the U.S.S.R. divide these belts along the southern boundary of typical tundra, i.e. they include both the northern tayga and the forest-tundra in the temperate-cold belt.*

In one way or another, only the thermal factor is taken into account in the differentiation of belts; any one belt, however, exhibits considerable contrasts in moisture-supply conditions. A belt does not constitute a single whole in the genetic sense. It must be concluded, therefore, that the differentiation of belts in regionalization is of no great scientific or practical importance. Belts can only be viewed as supplementary units at the very first stage of regionalization of global landforms, where they can be used to link together regional systems for individual continents or the major portions of such continents. In addition, the treatment of belts as groups of zones may possibly prove useful in the design of some applied regionalization systems, for agriculture in particular.

The principal category in zonal physical-geographic differentiation is a *landscape zone.* Considerable literature has been devoted to the zonal classification of the territory of the U.S.S.R. It has not been possible, however, to entirely exclude subjectivity in zone differentiation. This is largely due to the fact that landscape zones are still

---

* *Soil-Geographical Zoning of the U.S.S.R.* (1963) has been translated and published by the Israel Program for Scientific Translations.

established, as a rule, according to *ad hoc* partial characteristics. Additionally there exist difficulties which arise from the extremely gradual character of zonal change in space and from the existence of the broad transitional bands between zones. It is not accidental that the most serious differences in opinion emerge when the place of such transitional formations in the system of zonal classification is under discussion. It is very difficult, accordingly, to draw a clear distinction between zones and subzones.

The well-known system of L. S. Berg (Fig. 40), published during the 1930s, has served as the basic landscape-zonal classification of territory of the U.S.S.R. It has been used as a model by S. P. Suslov and other geographers. Here and there, however, there can be found more or less serious deviations from Berg's system. The following are only the most important. The forest-tundra, which Berg regarded only as a subzone of the tundra zone, has sometimes (in particular, in *The Natural-Historical Regionalization of the U.S.S.R.*) been treated as an independent zone. Yu. P. Parmuzin has argued recently in favour of a distinct forest-tundra and northern sparse-forest zone which would include a considerable proportion of the tayga zone, in the sense of L. S. Berg and other authors.

Another contentious problem is that of the independent status of the western Eurasian mixed-forest zone (this zone is better designated as 'subtayga'). L. S. Berg has conditionally grouped it together with the tayga into a single forest zone, and admitted the possibility that it may need to be differentiated as an independent zone. Today some geographers treat the subtayga as a zone, others treat it as a subzone. In *The Natural-Historical Regionalization of the U.S.S.R.* the southern part of the steppe zone is differentiated into a separate dry steppe zone. Very few geographers today would agree with this classification.

There is also disagreement about zonal boundaries. Thus F. N. Milkov, in contrast with most geographers, draws the eastern boundary of the mixed-forest zone not along the western slopes of the Urals but considerably further to the west, in the vicinity of the city of Gorki. The same author draws the boundary between the forest steppe and the steppe zones along the Russian plain, much further south than other geographers.[8]

There is now a tradition of landscape zone-differentiation on the basis of soil-geobotanical features. Taken independently of other physical-geographic characteristics, these features are not always reliable guides. Zones differentiated on the grounds that certain soil types or certain vegetation cover predominate, often do not coincide. Thus, for example, the chernozem zone does not completely coincide with the geobotanical steppe zone; the zone of chestnut soils likewise does not conform to any geobotanical zone. The question necessarily arises, which features should be given priority, soil or geobotanical? Neither solution can be regarded as final.

Some geographers consider that the best criteria for the plotting of zonal boundaries are heat and water supply indices. Hydrothermal factors and zonality indices have already been discussed in detail (see chapter 2), and the conclusion made that there is not direct inter-relationship between the hydrothermal coefficients and zonal boundaries. It should now be added that changes in hydrothermal conditions are very gradual; climatic and hydrological boundaries are extremely diffuse and provide no basis for a precise definition of landscape-zonal boundaries.

It is obvious that as long as the differentiation of landscape zones continues to be based on partial indices, the results cannot be either objective or consistent. In so far as landscape zones are geocomplexes, they must de differentiated with respect to sets of features reflecting the sum total of the major characteristics of zones. Simultaneously two aspects of the problem and, accordingly, two stages of the analysis, must be distinguished. First of all, the basic qualitative differences between zones must be defined and a general system of zones outlined taking into account the maximum possible number of indices but without a detailed consideration of zonal boundaries. Then, when such a system has been outlined in principle, the boundaries can be defined successively for each segment.

This analysis must begin from the study of the origin and the history of contemporary zones. Next, zonal variations in the climatic conditions must be examined without the restriction of a formal comparison of particular isolines (e.g. temperature, radiation balance, etc.) with the boundaries of the soil or the geobotanical zones, but by an analysis of all the radiation-circulation factors, taking account of the seasonal rhythm. The types of climatic conditions derived from these investigations must be reflected in the geochemical, geomorphological, soil and biogeographic processes. The establishment of this system of dependencies is the primary task.

Turning to a more detailed definition of zonal boundaries, account must be taken of the fact that the position of these boundaries depends in each case on a number of factors, and especially on topography and the bedrock which, as it were, transform the effect of a single zonal type of climatic condition. Examples of the effects of azonal factors on the position of zonal boundaries, in particular the boundary between the subtayga and the forest-steppe zones on the Russian plain, have already been presented. The soil and vegetation cover are always significant as indicators of zonal change; their role is particularly compelling where the geological-geomorphological conditions are uniform and the climate changes gradually producing extremely diffuse zonal boundaries.

The best method for the differentiation of landscape zones and the definition of their boundaries is to isolate individual landscapes and group them into zones (i.e. 'from below'). In this case the zonal boundaries coincide with the boundaries of individual concrete landscapes. In grouping landscapes into zones, reliance must be made on the fact that landscapes of particular types, comprising a series of zonal analogues (i.e. landscapes of desert or steppe types, etc.) predominate in each zone. There is a close link here between physical-geographic regionalization and landscape classification.

Landscapes of a single type, however, do not as a rule form a homogeneous area, but often exist as individual islands in adjacent zones. Thus, for example, the landscape of the Izhora plateau (the 'Silurian plateau') belongs to the eastern-European subtayga landscapes, yet it lies within the tayga zone. Isolated forest-steppe landscapes of central Siberia are also found within the tayga zone.

Thus the actual area covered by landscapes of a particular type does not necessarily coincide with the area of a single landscape zone. For this reason any zone may exhibit, in addition to the principal landscape types, isolated landscapes of other types.

To apply this method of zone differentiation, landscape maps and detailed landscape classifications are needed. Until these are available for the entire territory of the U.S.S.R., indirect methods must be relied on.

The most substantial qualitative changes in zonal phenomena occur along zonal boundaries. However, since each zone constitutes a broad band (usually 5° to 10° latitude) and zonal changes occur continuously and gradually, it is natural to expect the landscapes in the southern and the northern part of each zone to differ considerably from one another. In consequence of such continuous variation in the zonal characteristics of landscapes and the interpenetration of the features of adjacent zones, the latter can be differentiated into *landscape subzones.*

The criteria for the differentiation of subzones include the secondary zonal changes in landscape structure, i.e. the predominance of landscapes of a certain subtype. As regards the soils and the vegetation cover, subzones are usually characterized by the prevailing well-drained upland soil and vegetation subtypes. Subzones have long existed in landscape and branch-type regionalization although, as mentioned earlier, the taxonomic status of certain zonal units remains indistinct.

Theoretically it should be possible to distinguish within each zone three characteristic subzones: the central subzone in which the principal geographic characteristics of the given zone are most typically exhibited, and two marginal zones, northern and southern, the landscapes of which are characterized by many transitional features. Such three-unit systems, however, are not always realistic. They are most clearly manifested in zones with the most regular latitudinal extension and a substantial north–south spread, e.g. the Eurasian tundra zone, the Eurasian tayga zone and the eastern-Asian subtropical forest zone. Thus in the middle-tayga subzone the characteristic properties of the tayga zone are most marked; the northern-tayga subzone exhibits features indicating a transition to tundra (more strictly, forest-tundra) and the southern-tayga subzone a transition to subtayga.

Some landscape zones, however, (e.g. the forest-steppe and semi-desert) are inherently transitional and allow the distinction of only two subzones. There are considerable differences of opinion among geographers, soil scientists and geobotanists concerning the differentiation of the transitional zones and subzones. Some geobotanists, for example, do not differentiate the forest-steppe and the semidesert as independent zones, on the rather formal grounds that no specialized forest-steppe or semidesert types of vegetation exist and that both types of vegetation characteristic of adjacent zones coexist in each transitional zone. Accordingly, geobotanists regard the northern part of the forest-steppe as a separate forest (broad-leaved) zone and they attach the southern forest-steppe to the steppe zone as a grassland-steppe subzone. Similarly, they relate the northern part of the semi-desert (the 'desert-steppe' subzone) to the steppe zone, and the southern part (the 'steppe-desert' subzone) to the desert zone.[9]

These procedures, however, can hardly be justified even on purely geobotanical grounds. It is true, of course, that in the northern semi-desert the steppe vegetation prevails over desert vegetation, while the reverse holds true in the south. But the distinguishing characteristic of the plant cover in semidesert is the coexistence of the two types of plant communities and their characteristic mutual combinations ('complexes'). The physical-geographic conditions allow both types to develop normally. The same is true of the soils. On the ground it is virtually impossible to divide the steppe and desert communities and the semidesert soils in such a way that they would constitute two independent territories; this zone can only be divided into two parts according to the predominance of desert or steppe facies, i.e. in accordance with the quantitative ratio of one to the other. Such a ratio, however, constitutes a secondary level, i.e. it will characterize subzones and not zones. The independent status of the semidesert was noted towards the end of the nineteenth century by G. N. Vysotski who regarded it as a separate near-desert zone. In 1907 this zone was fully established by N. A. Dimo and B. A. Keller. Some geobotanists today also accept the semidesert as an independent zone.[10]

The forest-steppe also constitutes a separate landscape zone. The clear geographic individuality of the forest-steppe has been noted by V. V. Dokuchayev and his followers. The historical and physical-geographic conditions of this zone determine the distribution of the upland, well-drained localities of both forest and steppe communities, and the soils among which complex dynamic relationships operate.

The European subtayga is more correctly regarded not as a subzone of the tayga but as an independent zone. The tayga and subtayga differ considerably as regards their distribution. Whereas the former is most typical of the central parts of continents, the latter is found only in the coastal and inland transitional areas. As the subtayga wedges out towards the east, the tayga, by contrast, increases its area.

As new factual data are accumulated, the taxonomic validity of the various zonal classifications and zonal boundaries will continue to be reviewed. In particular, there is need to consider the possibility of differentiating subordinate landscape zones, or bands. These units have already been established in some cases by geobotanists and soil scientists, but they have not yet been studied by landscape scientists. It may be that landscape zones will have to be subdivided into bands when the qualitative changes of a zonal nature manifest themselves distinctly in only one or two components without affecting the others. For example, two subzones have been established in the forest-steppe zone on the Russian plain: the northern subzone where broad-leaved forests extend over larger areas than the grassland steppe, and the southern subzone where the grassland steppe predominates. With respect to the soils, each of these subzones can be subdivided into at least two further parts or bands: the northern into grey-forest soils and podsolized chernozem bands, and the southern into leached chernozems and typical chernozem bands. At the same time, these third-order zonal changes are not reflected in the climate or, to any degree, in the plant cover.

It should be added, however, that the disparities between the changes in various components may only be apparent. A good deal depends on the level of knowledge about the individual components. In the forest-steppe the soils are better known and better classified than the plant communities. In the tundra the situation is reversed; at least five zonal bands could be isolated on geobotanical grounds, whereas knowledge of the soils is not nearly so detailed. It is possible that a deeper and more co-ordinated study of soils, plants and other geographical components will produce a closer correlation among them.

Turning to the higher units of azonal differentiation, it must be noted that theoretically it is possible to speak of two categories of such units: the proper azonal categories associated with morphotectonic causes, and the sectional categories associated with longitudinal-climatic differentiation (meridional zonality or provinciality). The idea that continents should be divided into three meridional zones was first expressed in 1921 by V. L. Komarov, and revived by A. I. Yanputnin. This classification, however, is a somewhat idealized system and is not entirely applicable to concrete regionalization.

Yanputnin has shown that the physical-geographical differences between the meridional sections of continents are not clearly distinct at all latitudes, but only in the temperate and subtropical latitudes; the boundaries between them are closely correlated with the orographic properties of the continent and its overall size.[11] In investigating meridional zonality we are often compelled to differentiate not three but more sectors in a latitudinal belt. Thus in the Eurasian temperate latitudes, owing to the large size of the continent and the character of its major morphostructural properties, five sectors can be distinguished: western European, eastern European, western Siberian, eastern Siberian and far eastern. Two of these are near-oceanic, one distinctly intra-continental, and the remaining two transitional.

Thus longitudinal-climatic differentiation is influenced by the morphostructural characteristics of continents; for this reason it is pointless to contrast sectional differentiation or longitudinal provinciality with azonality proper, and accordingly to establish two independent categories of azonal regional units. The largest unit in the azonal classification which can be designated, after A. E. Grigoryev and A. I. Yanputnin, as the physical-geographic sector of a continent[12] is simultaneously characterized by a distinct morphostructural homogeneity and a specific character of atmospheric processes determined by the position of a given area in the system of the continental-oceanic migration of air masses.

Since both these factors operate jointly, each sector is distinguished by a characteristic series (spectrum) of latitudinal zones. If, for example, the eastern-Asian monsoon sector with its forest zonal series is compared with the central-Asian intra-continental sector with its desert-steppe spectrum of landscape zones, the fullest series of zones are found in the transitional sectors.

Almost every sector combines both plains and mountain regions. In most sectors the central position belongs to sizeable plains, plateaux or medium-high areas on ancient platforms; these constitute, as it were, a nucleus of the sector. Peripheral areas usually comprise an alpine frame consisting either of relatively young fold formations or 'rejuvenated' fold-fault ranges. Since sector boundaries generally coincide with mountain barriers, the sectorial structure of continents assumes a kind of cellular character. The mountains themselves, however, cannot be included within any one sector and they must be apportioned between two or more adjacent sectors with boundaries running along the crests of ranges. Thus the far-eastern mountains lie along the boundary of the far-eastern and eastern-Siberian sectors; the western slopes of the Urals must be included with the eastern-European sector and the eastern-Siberian with the western-Siberian sector. An even more complex picture prevails in the Great Caucasus and the Altai, both of which constitute landscape 'junctions' lying at the point of contact of several sectors.

The regionalization of alpine territories is extremely complex and requires additional treatment. At this point it should be noted that any independent mountain district or region constitutes a separate unit in physical-geographic regionalization. It follows that the differentiation of sectors contradicts the treatment of alpine territories as independent regional units. For this reason many geographers exclude sectors from their regionalization systems. This is entirely reasonable, but sector differentiation may play a useful supplementary role in the macro-regionalization of entire continents by providing the initial outline of azonal differentiation. Once detailed regionalization is undertaken, sectors lose their significance.

The most generally accepted level of azonal classification is the *district*, found in the majority of existing regionalization systems which deal with large territories.

The principal criteria for the definition of districts is as follows: (1) the unity of geostructure (old platforms, shields, orogenic regions) and an active history of recent tectonic movements; (2) common macrorelief features (broad lowland plains, plateaux, large mountain forms); (3) characteristic atmospheric processes and macroclimate associated with the hypsometric position and the degree of oceanic influence (the interaction between marine and continental air masses, their transformation conditions, degree of continentality and moisture); (4) the character of latitudinal zonality (the number of landscape zones, the peculiarities in their distribution and specific structural features); (5) the presence or absence of vertical stratification.

Some of the features also characterize sectors, but they achieve an appropriate concreteness in the definition of districts where alpine and lowland areas are properly distinguished. Districts are more distinct morphostructurally, are more homogeneous genetically, and their boundaries are sharper. One of the major criteria for sector differentiation—the structure of latitudinal zonality—manifests itself most clearly in lowland districts. In the mountains it is strongly transformed by altitude and the proximity of numerous mountain districts lying along the boundaries of different sectors.

Physical-geographic districts correspond only in general outline to large geostructural entities. A territory of a single district may include areas with a folded base if this base lies deep under the sediment and has little effect on the character of landscapes, or if it has been strongly transformed by recent tectonic movements which have unified structures of different age into an integrated whole. Thus the Russian plain lies on a Pre-Cambrian folded base to which, in the southern part, Hercynian structures are attached. Within the uniform Altai-Sayan alpine district, as well as the Kazakh fold district, both Caledonian and Hercynian structures are present, yet much later movements, primarily block movements, are of decisive importance for the differentiation of these districts.

About twenty districts have been differentiated for the territory of the U.S.S.R. Apart from the already mentioned Russian plain, which constitutes the nucleus of the eastern-European sector, examples include the Turan district and the western-Siberian district; mountain districts include the Urals, Crimean-Caucasus and Baykal districts etc.

The characteristic features of a district as a major unit in regionalization stand out most clearly when districts lying along approximately identical latitudes are compared. A good example is the Russian plain, the western-Siberian lowland and the central-Siberian plateau.

The Russian plain lies mainly on the Pre-Cambrian platform which underwent gentle uplift during the Neogene period and the Quaternary era. The western-Siberian lowland lies on Hercynian folded structures; the most recent movements show a downward tendency. The central-Siberian plateau comprises a Pre-Cambrian platform which has been considerably raised during the most recent geological epoch. The Russian plain lies, on the average, 170 m above sea-level, but its relief is fairly complex and is characterized by alternating uplands and depressions. Surface deposits are extremely diverse (mostly loose) and Quaternary glaciation has strongly affected landscape development. The western-Siberian lowland is characterized by a much lower hypsometric position (on the average, approximately 100 m above sea-level) and uniform relief; loose surface-deposits predominate. Quaternary glaciation here was not as significant as on the Russian plain; instead landscape formation was dominated by the accumulation of water-borne loose deposits. Central Siberia is much higher (on the average, approximately 500 m above sea-level), its relief is strongly dissected and locally alpine in character; dense bedrock of diverse structure predominates. Only a small part of central Siberia was affected by Quaternary glaciation.

The Russian plain lies closer to the Atlantic Ocean than either of the two districts; oceanic air masses play an important role in the development of its climate. It receives the maximum atmospheric moisture, while the continentality of its climate is minimal (the annual amplitude of mean temperatures lies between 20° and 35° C). The climate of the western-Siberian lowland is much more continental; in central Siberia, which lies at a maximum distance from moisture sources and coincides with the region of the Siberian winter pressure maximum, the amount of rainfall is only one-half that on the Russian plain, while the mean annual temperature variation reaches 40° to 50° C.

The differences in the history of landscapes, in the structure of the bedrock and in geographical position (longitudinal) are reflected in the character of permafrost, the degree of swampiness, plant cover, soils, etc. and, in general, in the characteristics of latitudinal zonality. Within the borders of the Russian plain permafrost is found only in the tundra, in western Siberia it extends to the northern tayga as well, while in central Siberia it occurs throughout. Marshes are widespread only in the northern half of the Russian plain, and in western Siberia they are virtually widespread throughout (they predominate in parts of the tayga); there are few marshes in central Siberia.

Zonality has the simplest character in central Siberia where the tayga constitutes the only zone. In western Siberia the zonal spectrum is broader and includes the tayga, subtayga, forest-steppe and steppe. On the Russian plain, within the corresponding latitudinal belt, the same zones occur, but they differ substantially in character. In particular, the subtayga broadens out significantly in a westerly direction and includes broad-leaved tree species. Dark-conifer species predominate in the tayga of the Russian plain and western Siberia, while light-conifer (larch) species predominate in central Siberia. Zonal boundaries along the Russian plain extend from the south-west to north-east, or from the north-west to south-east, and are greatly complicated by the effects of topography and surface outcrops. In western Siberia zones take the form of regular latitudinal bands, with extremely diffuse transitions; in central Siberia vertical stratification occurs in mountain areas.

Physical-geographic districts are divided, with respect to azonal features, into *regions*. The preceding section considered briefly the justification for this level of regional classification from the standpoint of the genetic principle, and it will be recalled that physical-geographic regions are regarded as units of districts which have become differentiated during the most recent stages of geological history (mainly during the Quaternary) under the action of azonal factors.

A region consolidates landscapes of common age and origin, as well as similar topography, surface deposits and hydrographic network. Unique orography and geographic position determine the homogeneity of landscapes within a region in the climatic sense as well. A landscape region may include landscapes of different zones. (Fig. 42) The zonal differences between landscapes belonging to a single region are lessened owing to their genetic unity. K. I. Gerenchuk and N. A. Solntsev drew attention to the fact that landscapes are often more closely interrelated within regions than within zones. Thus the steppe and forest-steppe landscapes in the High Transvolga region are in many respects closer to one another than, for example, the forest-steppe landscapes of the High Transvolga and the Volhyn-Podolye.

These circumstances enable us to differentiate the physical-geographic (landscape) regions as independent azonal complexes subordinate to a district, i.e. external to zonal classification. This does not, of course, provide sufficient grounds for the rejection of landscape zones but provides yet another proof that zonal and azonal classifications do not coincide.

A uniserial system of taxonomic units in regionalization

A uniserial system is one in which all taxonomic units of regionalization are distributed into a single subordination series. A relatively large number of variants of such a system have been proposed. The differences among them, however, are not substantial. We shall consider a few of the most notable examples.

A. A. Grigoryev proposed the following system of units based on the alternation of the major zonal and azonal factors: belt-sector-zone and subzone-province-landscape (rayon).

A slightly different schema was offered by V. B. Sochava: belt-district-zone and subzone-province-landscape (okrug).

F. N. Milkov offered the following: belt-district-zone-province-subzone(bands)-rayon. A similar schema was proposed by Yu. P. Parmuzin and a number of other geographers. The Inter-University Conference on Regionalization recommended the following system: district-zone-province-subzone-okrug-rayon.

Disregarding certain differences in detail, it is clear that these series are based on a common principle involving alternative utilization of

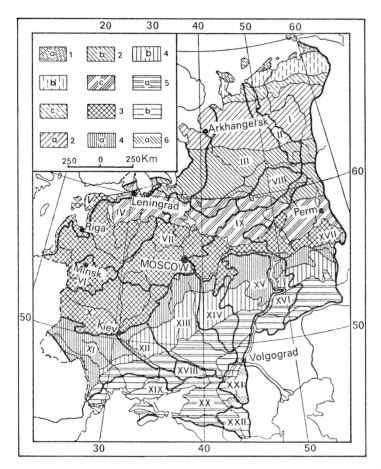

Fig. 42 Physical-geographic regionalization of the Russian plain (after A. G. Isachenko)

Landscape zones and subzones: *1*, tundra (*a*, moss-lichen; *b*, scrub; *c*, forest tundra); *2*, tayga (*a*, northern; *b*, middle; *c*, southern); *3*, subtayga; *4*, forest steppe (*a*, northern; *b*, southern); *5*, steppe (*a*, northern; *b*, southern); *6*, semidesert (*a*, northern). *Landscape regions*: *I*, Pechora; *II*, Timan; *III*, Dvina-Mezen; *IV*, north-western; *V*, Baltic; *VI*, upper Neman; *VII*, upper Volga; *VIII*, Severnyye Uvaly; *IX*, Kama-Myershchera; *X*, Polesye; *XI*, Volyn-Podolye; *XII*, Dnepr; *XIII,* central Russian; *XIV*, Oka-Don; *XV*, Volga; *XVI*, lower Transvolga; *XVII*, High Transvolga; *XVIII*, Donets; *XIX*, Black Sea; *XX*, Azov Sea; *XXI*, Yergenin; *XXII*, Stavropol.

zonal and azonal features for the differentiation of successive taxonomic units. Thus the district or the sector constitute azonal subdivisions of a belt, the zone constitutes a zonal unit within a district
i.e. subordinate to the district, and the province constitutes an azonal
unit subordinate to the zone, etc.

Many geographers consider, however, that the alternation of zonal
and azonal features is entirely arbitrary and does not reflect the natural
interrelationships between geocomplexes. These taxonomic systems
were also criticized by Yu. K. Yefremov and other geographers. A
landscape zone cannot be regarded as arising from some secondary
causes within individual physical-geographic districts (as maintained
in the uniserial system), since zones reflect the general law of zonality
and not the laws characteristic of a given district. Subzones are directly
associated with zones, being a natural result of the latter's differentiation and not of the differentiation of provinces.

Nevertheless, some geographers (e.g. F. N. Milkov, V. I. Prokayev
and Yu. P. Parmuzin) raised the alternation of zonal and azonal
features to the rank of the major principle for the taxonomic system
of regionalization, maintaining that this leads to the unification of
zonality and azonality at every level of the taxonomic ladder. Such
unity is achieved, however, at the expense of the internal unity of
many of the higher categories in regional classification. It is impossible
to design a single series of units without simultaneously disturbing the
genetic units of such complexes as zones, subzones and regions. In a
uniserial system the landscape zone entirely disappears and is fragmented into segments subordinate to districts. There is no place in
such systems for the Eurasian tayga zone or the Eurasian steppe zone.
In their place we find a whole series of independent zonal units within
each district (the tayga zone of the Baltic shield, the tayga zone of the
Russian plain, etc.). Such units are, of course, quite realistic and should
enter a taxonomic system of regionalization, yet this should not
prevent the consideration of a zone within continental boundaries as
an independent unit of a higher rank.

In terms of the uniserial system, parts of the steppe zone, e.g. parts
of the western-Siberian lowland and of the Kazakh fold district, must
be regarded as two different zones subordinate to the corresponding
districts. Both these zones, however, form a single area comprising in
turn a part of the Eurasian steppe zone, a still larger regional unit.

Some geographers maintain that the differentiation of the entire
tayga as a regionalization unit is of no theoretical or economic interest.
This, however, raises a question. Does the differentiation of belts
(temperate, subtropical, etc.) forming part of most uniserial systems
have any greater theoretical or practical significance than the differen-

tiation of zones? Other authors accept the existence of landscape zones in a broad sense, but nevertheless exclude them from taxonomic systems where their introduction within a single series would disrupt the units of districts and other azonal units.

Subzones create an even more complex situation. According to F. N. Milkov, for example, on the Russian plain alone eight northern forest-steppe subzones and an equal number of central and southern subzones need to be differentiated. It appears that there can be no single subzones in the northern forest-steppe, or even in the northern forest-steppe of the Russian plain, since purely formal criteria (the 'alternation principle') require the subzone to be subordinate to a province. In nature, on the other hand, subzones are so closely associated with zones that it is often difficult to distinguish them from one another; examples of this were provided in the preceding section. It follows that a directly subordinate relationship exists only between the subzone and the zone. In a uniserial system the status of the eastern-European subtayga will depend on whether it is regarded as a zone or a subzone. In the former case it will constitute an independent unit divided into a number of provinces, and in the second it will appear as a number of subzones isolated within the boundaries of provinces. In reality, however, neither the integrity nor the independence of the eastern-European subtayga can be affected by its treatment as a zone or subzone.

As noted earlier, landscape regions are excluded from the uniserial system of regionalization; they are actually distributed among the zones. This exclusion is entirely unjustified, especially since no one has any doubts that the more complex and genetically heterogeneous azonal complexes, i.e. districts, must be included in the system. It follows that the uniserial system is purely a convenient device. In reality the taxonomic units arise from the intersection of two independent series: zonal and azonal. Uniserial systems always give priority to either of the two major factors in physical-geographic differentiation in accordance with each geographer's subjective opinion. Some give priority to the belt, thus, as it were, stressing the precedence of zonal origin; others give priority to districts or even sectors, giving precedence to azonal factors.

The desire to include zonal and azonal units in a single hierarchy inevitably leads to numerous errors and distortions. Thus if the belt constitutes a unit of the first rank, then the sector or district must constitute the next level. Yet neither sectors nor districts, being units the existence of which is due to zonal causes, ever conform without overlapping with the boundaries of any given belt.

The eastern-Asian monsoon sector extends over at least three different belts (e.g. temperate, subtropical and tropical). Many districts also cross the boundaries of various belts. Particularly complex in this regard are alpine districts. To resolve these problems the differentiation of belts must be abandoned altogether or augmented considerably.

Alpine districts are often treated separately, independently of belts, or are assigned to belts arbitrarily. Some supporters of the alternation principle refuse even to differentiate the subzones, or contradict their own criteria by treating subzones as directly subordinate to zones. As a result of these deviations and inaccuracies many geographers employ in practice a system of units which has little in common with a theoretically consistent hierarchy.[13]

Thus the traditional method of representing a system of regionalization units as a single series cannot be regarded as the only possible solution to the complex problem of accounting for zonal and azonal factors in regionalization.

In favour of the uniserial system there is only our familiarity with it and its considerable utility. A uniform system underlies every description; there is no need to produce simultaneous descriptions of two regional units of a different type; they can be described in sequence. Sequential description of territorial units, however, has nothing to do with the principles of regionalization but has to do with the methodology of regional description in physical geography. It would hardly be in order to subordinate the development of taxonomic systems in regionalization to the methodology of description.

The experience of many scientific disciplines shows that classifications hardly ever conform to single subordination series. We need only recall the periodic table of chemical elements compiled by D. I. Mendeleyev, and the phylogenetic classification of organisms which takes the form of a branching tree, or the ecological-phytocenotic series of plant associations compiled by V. N. Sukachev. Obviously classification systems for different natural phenomena cannot follow a single recipe.

The biserial system of taxonomic units in regionalization

The biserial system is based on an objective treatment of both the primary series—zonal and azonal—which merge in the landscape and which are associated on intermediate levels by derived or linking units.

The idea of biserial regionalization is not new. One can find it in the work of R. I. Abolin, going as far back as 1914. L. S. Berg's well-known 1913 paper on the landscape and morphological regionalization of Siberia and Turkestan can also be regarded, up to a point, as an example of a disparate regionalization by zonal and azonal factors. In 1919 P. N. Krylov published a geobotanical regionalization of Siberia which provides an explicit account of two series of regional units. The first includes latitudinal-zonal subdivisions, and the second azonal subdivisions based mainly on surface character.

D. L. Armand, A. G. Isachenko and other geographers have supported the application of the biserial principle to physical-geographic regionalization. In recent years this principle has been steadily gaining acceptance. A. A. Grigoryev, for example, has accepted the need for the establishment of azonal (interzonal) districts and regions embracing parts of certain belts and zones.[14] V. I. Prokayev has concluded in a recent paper that there exist zonal and azonal physical-geographic regionalizations which do not coincide, but the intersection of which gives rise to a third, functional order of units.[15] The zonal and azonal systems of units are treated independently of one another by A. M. Ryabchikov, E. N. Lukashova and Yu. K. Yefremova. I. P. Gerasimov has postulated two series of taxonomic units in soil regionalization.

All the existing physical-geographic regionalizations of the U.S.S.R. are in fact based on the biserial system, i.e. in accordance with the principle of *independent* differentiation of zonal and azonal units; this is easily seen when any existing regionalization map is examined. The apparent use of a single series is merely due to the need to arrange regions according to some explicit convention.

An example of such a system is the natural-historical regionalization of the U.S.S.R. published in 1947 by the U.S.S.R. Academy of Sciences. (Fig. 43) The authors of that regionalization maintained that 'the highest and most general unit' is the natural-historical zone, but their description is in terms of districts. In actual fact, zones and districts were differentiated independently of one another and are entirely unrelated. This is clearly brought out by a map on which the zones are depicted as continuous bands (on the original map each zone is printed in a different colour) which intersect the boundaries of numerous districts. At the same time districts do not conform with zonal boundaries, and this too is clearly shown on the map.

Parts of the same zone lying within individual districts are regarded in natural-historical regionalization as provinces. In many cases, however, provinces are regarded as smaller units and their criteria are obscure. The absence of subzones must also be regarded as a defect.

In the physical-geographic regionalization of the Asian territory of the Soviet Union produced by S. P. Suslov (Fig. 41) the zonal and azonal units are not subordinate to one another. On his maps, zones and subzones are shown as bands intersecting the boundaries of districts and regions.

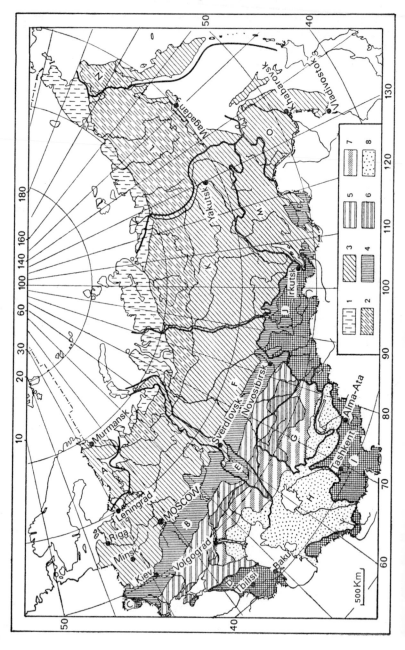

The well-known inconsistency of S. P. Suslov's system resides in his failure to extend the boundaries of zones and subzones to alpine regions. The same criticism can also be directed at a system of natural-historical regionalization where only the low-level mountain districts are divided into zones. S. P. Suslov's effort, however, has a number of advantages over natural-historical regionalization. The principal merit of Suslov's system is his deeper and more detailed physical-geographic description of regions. In addition this system is distinguished by a more detailed classification of terrain, with respect to zonality.

The most recent regionalization of the entire territory of the Soviet Union is that of G. D. Richter.[16] This system is based on only two taxonomic units: a district and a province. Districts are differentiated in accordance with the generally accepted criteria for this category; in regard to provinces, their criteria are somewhat lacking in precision. The author stresses that 'each province is contained within a zone'. The zones themselves, however, are not among the taxonomic units. Nevertheless Richter groups provinces into 'zonal and azonal groupings'. It is quite obvious that the zonal groups of provinces depicted on Richter's map by a different shading constitute nothing other than the standard (continuous) landscape zones, and also partly subzones. Thus, on purely formal grounds the author failed to introduce zones into his system of units, since they could not be regarded as subordinate to districts and, consequently, could not form a series. In actual fact zones do exist in that system quite independently of districts and provinces. In other words, here too we are concerned with biserial regionalization. As regards the 'azonal groups of provinces', these do not constitute regional units of any kind and appear to be merely areas of similar landform: lowland, plateau and highland, low- medium- and high-alpine districts. Similar examples could be adduced from investigations into branch regionalization, in particular hydrological, geobotanical and soil regionalization.

Thus all uniserial regionalization systems which do not distinguish zones within continental boundaries do in fact exhibit zones in an

---

Fig. 43 Schema for a natural-historic regionalization of the U.S.S.R. (Council for the Study of Productive Resources, U.S.S.R. Academy of Sciences, 1947) *Natural-historic zones*: *1*, tundra; *2*, forest tundra; *3*, forest; *4*, forest steppe; *5*, steppe; *6*, arid steppe; *7*, desert steppe; *8*, desert. *Natural-historic districts*: *A*, Baltic crystalline shield; *B*, eastern-European plain; *C*, Carpathians; *D*, Crimean-Caucasian alpine district; *E*, Urals; *F*, western-Siberian lowlands; *G*, Kazakh-fold district; *H*, Turan lowlands; *I*, Central-Asian alpine district; *J*, Sayan-Altai mountain district; *K*, central-Siberian elevated plain; *L*, Kolyma-Okhotsk alpine-lowland district; *M*, Transbaykal alpine district; *N*, Koryak-Kamchatka alpine district; *O*, far-eastern alpine-lowland district.

indirect way i.e. outside the adopted system of units. The same is true of the physical-geographic regions which implicitly enter the system and comprise the provinces of the same name in adjacent zones (e.g. one region consists of the Volga province of the forest-steppe zone and the Volga province of the steppe zone, or the Caspian province of the semidesert zone and the Caspian province of the desert zone).

It is not sufficient, however, to merely acknowledge the independent character of the zonal and azonal series in regionalization. As noted earlier, the principal task is to interrelate them without simultaneously destroying the internal unity of each system.

The most important zonal-azonal unit common to both series is a *landscape*. The landscape forms simultaneously a part of the landscape district and of a zone (or subzone), i.e. it is doubly dependent. The position of a landscape in a regional classification is determined, as it were, by two co-ordinates: zonal and azonal.

The physical-geographic regionalization of a territory is not always carried down to the level of landscapes. Very often it terminates at higher levels, depending on the purpose or availability of data. It is important, therefore, to be able to relate zonal with azonal classification on any level of regionalization so as to classify every territory accurately with respect to each series.

It is known that every zone or subzone, while maintaining its territorial unity, is transformed and assumes specific features inside each district or region. We can likewise say that individual areas of a district or a region exhibit characteristic properties depending on their position in the zonal classification of the earth's surface. This enables us to establish certain *derived* or linking units which combine the zonal series with the azonal and make it possible to pass from one series to the other along any level of regionalization.

There are two major connecting units: a section of a zone within the boundaries of a district, and a section of a zone within the boundaries of a region. If a zone subdivides into subzones, then two additional units must be provided: a subzone inside a district, and a subzone inside a region. The whole system is illustrated in figure 44.

Sections of zones within district boundaries generally remain undesignated; in uniserial systems they are simply called zones, and this may lead to confusion, since groups of such zones often comprising a single band extending across several districts or even across an entire continent are also called zones (i.e. zones in the sense of L. S. Berg and other geographers). To avoid employing a new designation, those parts of zones lying within a single district are referred to as *zones in the narrow sense*, by contrast with standard zones, or zones in the broad sense. A zone in the narrow sense can be easily distinguished

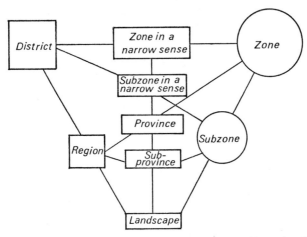

Fig. 44   System of taxonomic units in physical-geographic regionalization

from a standard zone by assigning it an appropriate name. Included with it, as a rule, is the name of the corresponding district (e.g. the forest zone of the Russian plain, the tayga zone of central Siberia, etc.) Such a designation reflects the *dual* subordination of this taxonomic unit; it is simultaneously a part of a district and a part of a single continental zone (in the uniserial system, a zone is formally subordinate only to a district).

The part of a zone lying inside a single landscape region (or, in other words, a section of a region within a zone) is usually called a landscape *province*, and the corresponding section of a subzone may be called a *subprovince*. Great importance attaches to these units in regionalization, since in cases where regionalization stops at a level short of distinguishing landscapes, provinces or subprovinces often constitute the terminal levels associating the zonal and azonal series. As mentioned already, landscape provinces enter into all uniserial systems, but are regarded in such systems only as units subordinate to zones or subzones, being geomorphologically or climatically distinct zone or subzone sections. Such a definition of a province is rather one-sided. In biserial systems a landscape province is regarded as part of a single azonal region within the boundaries of a given zone. This expresses the genetic dependence of provinces and their dual subordination. Individual sections of a landscape region lying in various zones or subzones are genetically uniform, but their development proceeds in non-uniform zonal conditions. This is conveyed by the names of provinces and subprovinces, which, as in other derived units,

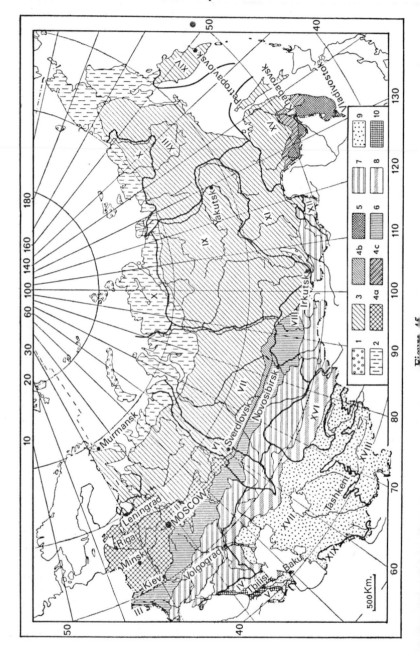

Figure 45

have a two-part character, e.g. the Pechora tayga province, the Pechora middle-tayga subprovince (these may be altered to the tayga province of the Pechora region or the middle-tayga subprovince of the Pechora region).

When a landscape region lies entirely within a single zone the province coincides areally with the region. The Black Sea region, for example, lies entirely within the steppe zone and, accordingly, constitutes its province. (Fig. 42)

A physical-geographic regionalization should always terminate in the differentiation of the connecting units, which also include a landscape. The connecting units do not constitute an independent series; they are mutually interconnected via the zonal and azonal factors, i.e. are units with *double subordination*. Figure 44 clearly illustrates the schema.

Three levels of regionalization can be distinguished, depending on the degree of detail introduced. The first level consists of the higher levels of the zonal and azonal series, i.e. zones and districts, the intersection of which yields the connecting units of upper rank—zones in the narrow sense. Figure 45 illustrates a physical-geographic regionalization of the U.S.S.R. performed at this level.

The second level of regionalization must include the successive stages (regions and subzones) and terminates in provinces and subprovinces. An example of regionalization on the second level is the classification of the Russian plain shown in figure 42. Finally, the third level of regionalization terminates in the differentiation of landscapes (rayons). An example is provided by the landscape regionalization of the Pechora tayga province in figure 46.

Occasionally intermediate units (designated as okrug) are established between the landscape (landscape rayon) and a province or

Fig. 45 Schema for a first-order physical-geographic regionalization of the U.S.S.R. (after A. G. Isachenko)
Landscape zones: *1*, arctic; *2*, tundra; *3*, tayga; *4*, subtayga (*4a*, European; *4b*, western Siberian; *4c*, far eastern); *5*, broad-leaved forest (far eastern); *6*, forest steppe; *7*, steppe; *8*, semidesert; *9*, desert; *10*, mediterranean (including wet subtropical barrier-foot landscapes and transitional Transcaucasian arid forest landscapes). *Landscape districts*: *I*, Fenno-Scandia (Baltic shield); *II*, Russian plain; *III*, Carpathian mountains; *IV*, Crimean-Caucasian mountain district; *V*, Armyan highlands; *VI*, Urals; *VII*, western-Siberian low-lying plain; *VIII*, Altai-Sayan alpine district; *IX*, central-Siberian district; *X*, northern-Siberian district; *XI*, Baykal alpine district; *XII*, Mongolian-Daurian district; *XIII*, north-eastern Siberian alpine district; *XIV*, Kamchatkan-Kurilian district; *XV*, Amur-Maritime district; *XVI*, eastern Kazakhstan (Kazakh-fold) district; *XVII*, Turan low-lying plain; *XVIII*, central-Asian alpine district; *XIX*, Turkmenian-Khorasanian alpine district.

252   *Landscape Science*

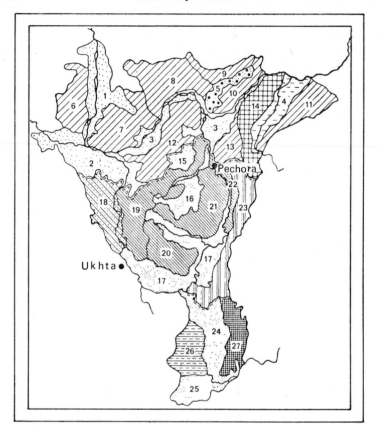

Fig. 46   Third-order physical-geographic regionalization of the Pechora tayga
province with landscapes numbered (after A. G. Isachenko)
Marginal northern-tayga subprovince: *1*, lower Pechora valley; *2*, Tsil'ma-
Izhma; *3*, Pechora-Usa valley (frontal-apron alluvial plains with large areas of
lowland and upland moors, sparse pine stands on podsolic bog, illuvial-humus
soils and illuvial-humus podsols, with flood meadows); *4*, Inta (flat, clayey-
sandy lacustrine plain with hummocky turf-peat and sphagnum bogs); *5*, Khata-
yakha (low-lying lacustrine-alluvial plain with dwarf birch, sparse spruce stands
and marshes, with light tundra-bog and podsolic-boggy soils); *6*, Tobysha;
7, Sos'vina; *8*, Laya-Yur'yakha; *9*, Kolva-Khatayakha; *10*, Usa, right bank; *11*,
Lemva; *12*, Pechora-Lyzha; *13*, Synya-Pechora (morainic ridgy plains with spruce,
mainly boggy and sparse forests on podsolic-boggy, loamy and sandy-loam
soils, with stretches of bog and tundra); *14*, Kryazh-Chernysheva (highland
plateau with karst). *Northern-tayga proper subprovince*: *15*, Lyzha; *16*, Kozh-
vina; *17*, Lem'yu-Pechora Nibel'ya (flat frontal-apron lacustrine-alluvial plains
with swampy pine stands on podsolic-boggy illuvial-humus soils and sphagnum
swamps); *18*, Timan-Izhma (morainic frontal-apron well-drained plain with

subprovince, but the distinction is not sufficiently clear. In regionalization the possibility of introducing, whenever necessary, supplementary (intermediate) taxonomic categories must always be retained. There is hardly any justification, however, to introduce yet another obligatory level above the landscape.

One example of an intermediate unit of regionalization is the *subregion*. In the north-western region of the Russian plain the Valdai, Ilmen and Baltic-Ladoga subregions can clearly be distinguished. The latter should, perhaps, be included with the Baltic region, as has been done in figure 42. Parts of the subregions within a single subzone can be designated as *okrug* (e.g. the Valdai southern-tayga okrug, the Valdai middle-tayga okrug). In many regions, however, the subregions, and therefore the okrugs as well, are not differentiated. These units can be regarded, therefore, as supplementary.

Thus the biserial system expresses most fully the regular character of geographic differentiation. While maintaining the unity of all the primary zonal and azonal classifications, it also enables the combination of the two series at any level of regionalization.

In one respect the biserial system must be regarded as inadequate: it is impossible to describe simultaneously two series of regions, especially if they are connected by derived units. This criticism, however, is not well founded. It concerns essentially the well-known imperfections of the descriptive method. Geographers also employ a 'second language', i.e. maps, which convey spatial relationships much better than text and supply the principal means by which these relationships are established. The representation of the biserial system on a map is not only very easy but virtually automatic. The map, as has already been remarked, constitutes the visual proof that this system is well founded. The simplest and, simultaneously, the most effective method of compiling regionalization maps is one in which units of a single series (predominantly zonal) are depicted in red, whilst units of

---

spruce and pine forest on gleyey-podsolic soils and humus-ferruginous podsols); *19*, Sebys-Kozhvina; *20*, Vel-Ayyuvina; *21*, Soplesa; *22*, Pechora, right bank (morainic plains, locally elevated, with monticulate relief, spruce forests, largely swampy); *23*, Uralian (poorly dissected swampy plain with a thin cover of dusty loams, with haircap-moss spruce stands on podsolic-boggy soils and bog stretches). *Middle-tayga subprovince*: *24*, upper Pechora; *25*, Berezovka (lowlying frontal-apron lacustrine plains with swampy pine stands on podsolic-boggy illuvial-humus soils, with sphagnum swamps); *26*, Mylva (flat plain with dusty loams in interfluves and sands in glacial meltwater drainage lines, with spruce and pine forest, mainly swampy); *27*, Uralian (poorly dissected plain with a thin cover of loams, spruce and spruce-fir forests on strongly podsolized and podsolic-boggy soils and stretches of sphagnum swamps).

the other series (azonal) are shown by different kinds of shading. Zonal and azonal boundaries are depicted by lines of different type and colour.

Descriptions of regions, like all descriptions, are always uniserial. Events or phenomena which occur simultaneously must necessarily be described sequentially. In regional descriptions therefore, we are faced with the problem of firstly outlining the most appropriate sequence to account for our material. This question concerns the method of regionalization, and it shall be returned to later. The advantages of the biserial system are particularly distinct in the regionalization of alpine areas, to which we now turn.

Special features of the physical-geographic regionalization of alpine territories

The physical-geographic regionalization of mountain regions is based on the same principle which operates in the regionalization of the plains. Mountain areas, however, have always created considerable difficulty. The individual character of mountain landscapes, the marked vertical differentiation and territorial uniqueness of mountain ranges, force us to regard them as independent regional units. On the other hand, the opposing slopes of many alpine systems differ so strongly that they must be distinguished from one another. At the same time many diverse genetic, climatic, biogeographic and hydrographic bonds between the landscapes on alpine slopes and the adjacent plains enable us to regard the individual parts of the alpine systems, together with the adjacent plains, as single physical-geographic entities. These bonds are clearly reflected in the structure of vertical stratification long established for the Alps, the Crimean mountains, the Great Caucasus, and other mountain ranges.

Accordingly, the watershed crests of ranges often constitute the most important physical-geographic boundaries. They frequently coincide with the boundaries of landscape zones and sectors. When boundaries are traced along the crest of a range, however, a territory which constitutes an integrated whole as regards orography, structure and genesis is subdivided into its constituent units. Some geographers see these procedures as mutually exclusive and contradictory. A. Hettner drew the ultimate conclusion that regionalization as such is an entirely subjective exercise. Nevertheless this contradiction could not be resolved, only because most geographers insisted on arranging regionalization categories into a single subordination series. That approach obviously cannot preserve the internal unity of large mountain regions while simultaneously combining the classification of alpine regions with that of the plains. The problem was always resolved in a one-sided manner: mountains were excluded from the general system of zonal classification and were regarded as purely azonal formations (examples are provided by the systems of L. S. Berg and S. P. Suslov discussed earlier), while the internal regionalization of the alpine regions themselves was approached either from the point of view of their orography or their vertical stratification.

Thus in L. S. Berg's well-known work devoted to the landscape zones of the U.S.S.R., the natural conditions inside each mountain region are described in terms of vertical strata. S. P. Suslov describes relief and the geological structure of alpine regions separately, with respect to orography or morpho-tectonic classification (e.g. in the Transbaykal and Baykal in terms of 'geomorphological subregions', such as the western Transbaykal, eastern Transbaykal, near-Baykal, etc.), while soils and the vegetation are classified in terms of vertical strata specific to a given mountain region and without any account of the variation in their structure.

This approach does not address itself entirely to the problems of regionalization. Vertical strata, which often consist of diverse sections, cannot be regarded as regional geocomplexes and their place is outside the taxonomic regionalization system; they cannot be included with such categories as the zone, district, province, rayon, etc. In recent years, therefore, the majority of geographers (e.g. I. S. Shchukin, V. M. Chetyrkin, V. B. Sochava and N. A. Gvozdetskiy) have come to regard the vertical strata (more correctly, parts of such strata) as morphological subdivisions of an alpine landscape region, i.e. essentially as units in intra-landscape classification.

Many experts have stressed the necessity of accounting for the *structure** of vertical stratification in physical-geographic regionalization. Attempts to base regionalization on differences in the structure of the vertical strata were made by geobotanists E. V. Shiffers, E. M. Lavrenko and V. B. Sochava, soil scientists A. N. Rozanov, V. M. Fridland and N. N. Rozov, and physical-geographers V. M. Chetyrkin, V. B. Sochava and V. I. Prokayev. Since many types and variants of vertical stratification reflect the zonal and provincial (longitudinal) differentiation of mountains, the structure of vertical stratification attains the status of the principal criterion in physical-geographic regionalization. This does not mean, however, that the regionalization of mountains must be based exclusively on this criterion.

The combined zonal and azonal approach remains the fundamental principle in the regionalization of mountains and plains alike. The

---

* The form and distribution of vertical strata. See chapter 2, p. 81.

problem of the regionalization of mountain areas is to depict mountains as individual regions, thus emphasizing their geographic uniqueness and simultaneously to express their diverse territorial bond with lowland landscapes.

This problem can be resolved if it is approached in terms of the biserial regionalization system which avoids the narrow azonal approach to the physical-geographic classification of mountains, takes into account the effect of zonal factors, and combines in a single system the regionalization of mountains and plains. Experience has shown that the same taxonomic categories can be used for the physical-geographic regionalization of both mountains and plains. In view, however, of the complexity of mountain landscapes, certain additional levels of classification need to be introduced.

Every mountain range constitutes an independent *azonal* unit of regionalization. The taxonomic significance of such a unit may differ depending on its size and complexity. Very large alpine territories with complex orographic and tectonic structure, which extend over many different zones or sectors and are therefore characterized by a great many longitudinal variants of vertical stratification, are regarded as separate physical-geographic districts. Examples of physical-geographic alpine districts are the Carpathians, the Urals, the Great Caucasus, the Altai-Sayan highlands and the alpine district of north-eastern Siberia. (Fig. 45)

Parts of alpine districts clearly defined in the orographic and tectonic aspects (e.g. the Upper-Kolyma highlands, Cherski Range, Yukagir plateau in north-eastern Siberia, eastern Sayan, the Tuvin basin, Kuznetski Ala-tau in the Altai-Sayan alpine district), as well as independent mountain ranges, are relatively simple as regards their orography and vertical stratification (e.g. the Crimean mountains, the Putoran and Byrrang mountains, etc.) and constitute physical-geographic regions.

Finally, certain small 'island' alpine elevations on plains (e.g. small anticlinal ranges, volcanic cones etc.), if characterized by a uniform vertical stratification, constitute individual mountain *okrugs*, *rayons*, or *landscapes*. Among them are the Khibins and other mountain areas of the Kola Peninsula, the Balkhans, the residual hills of the Kyzyl Kums.

The intra-range basins constitute independent (azonal) units of alpine districts and regions. The largest among them, for example the Tuvin or Minusin basins, have the status of physical-geographic regions and, just like the lowland regions, divide into rayons (landscapes). Inter-montane basins with a simpler landscape structure (e.g. the Gorii and Akhaltsikh basins in the Caucasus, a series of graben-depressions in the Stanov mountains) must be regarded as independent rayons (landscapes).

A much more difficult problem concerns the status in a system of large regional subunits of inter-montane basins open in the direction of the sea, or of the adjacent platform-plain districts with which they are directly connected (e.g. the Kurin and Rion basins, the Kuznets depression). In a taxonomic sense they are typical physical-geographic regions; the only question is whether they should be treated as mountain or plain regions. As regards their landscapes, they are undoubtedly plains. Such landscapes, however, are genetically associated with the surrounding mountains; they result from the tectonic differentiation of a single entity. Moreover, their physical-geographic characteristics are substantially determined by the effect of the alpine framework. Specifically, they exhibit the barrier foot and barrier shadow phenomena and frequent zone inversions, which were discussed in chapter 2.

The sharp contrast between two adjacent Transcaucasian plains (e.g. the Rion plain with its wet subtropical forests and the Kurin plain with its semidesert and desert landscapes) is explained by their relative position with respect to the mountain ranges. Such examples can be multiplied *ad infinitum*. They point to the need to unify inter-montane plains of the open type with the associated alpine elevations into a single regional complex, even though their landscapes may be typologically very diverse and therefore occupy different positions in the taxonomic system.

It follows that the classification of districts into alpine and lowlands is quite arbitrary. In many districts normally regarded as lowland, individual mountain landscapes or even regions are found (e.g. the Turan plain district includes the mountain regions of Mangyshlak and Tuarkyr as well as a number of alpine okrugs and landscapes). Moreover, a considerable proportion of alpine districts include lowland regions and landscapes. There also exist 'mixed' mountain-lowland districts with roughly equal proportions of mountains and plains areas. For example, in the Amur district, which is regarded as alpine, alongside the mountain regions (Sikhote-Alin, Burein, etc.) are found broad lowland (inter-montane) regions (Amuro-Zei and Amuro-Sungari-Ussuri etc.)

The problem of the zonal classification of mountains is based on the investigation of types of vertical stratification. The latitudinal-zonal position of an alpine region is reflected in the corresponding types of stratification. It follows that parts of alpine districts or regions characterized by a common vertical stratification belong to a single *landscape zone.*

It should be borne in mind, however, that it is not easy to define zonal boundaries in mountain regions. The successive alternation of zones is more distinct on ranges intersecting zones from north to south and ranges with a relatively simple orographic structure, such as the Urals. Even in such cases, however, the zonal boundaries level out in altitude and must be drawn more or less along the crests. In mountains, the combination of zonal with longitudinal phenomena may assume great complexity, depending on orographic structure. Owing to orographic factors, the spatial relationships between zones in alpine districts are subject to fairly sharp changes. Zonal boundaries are not in the same positions in which they would have been if the alpine ranges were replaced with lowland plains. Even along the plains, relief induces displacements in zonal boundaries; these displacements are more acute in alpine regions.

The presence of high-altitude crests and the contrasts between different exposures lead to a situation in which the distribution of landscape zones in alpine regions is strongly subordinated to the orographic structure of the region. This situation facilitates to some degree the definition of zonal boundaries. In high-altitude ranges, especially those in which the strike coincides closely with the general trend of the landscape zones, the boundaries of these zones often run along the crests. An example of such a zonal boundary is the crest of the Great Caucasus. Admittedly the fact that the zonal differences between opposing slopes are often smoothed out with increasing proximity to the axial section of high-altitude ranges must be taken into consideration. The vertical-stratification series of opposing slopes terminate in the common peak stratum, which has to be divided between two different zones or arbitrarily included with one of them. This situation should not hinder the zonal classifications of mountain ranges. It will be recalled that even over the plains, landscape zone boundaries are rather diffuse and cannot be plotted objectively.

Complex orographic dissection and especially the presence of opposite aspects, which complicate the diversity of vertical stratification, create certain difficulties for the definition of landscape zones. It should be borne in mind that zonal regionalization of alpine areas has only just begun and very little is known about this subject. A considerable effort awaits geographers in this area, especially as regards the structure of vertical stratification. In mountain districts and regions, landscape *subzones* can also be defined. Furthermore, everything said above about zonal boundaries applies equally well to the subzone.

The association between zonal and azonal classification in alpine regions is similar to that obtaining on the plain: derived regional categories are differentiated. In the case of alpine regions, zones and *subzones in the narrow sense* can be differentiated (e.g. the tayga zone of the Amur mountain district, the southern-tayga subzone of the same district). However, in mountain regions with their orographic contrasts and sequences of ranges and inter-montane basins, the differentiation of landscape *provinces* is far more important. Landscape provinces are defined as a part of the mountain region characterized by a single type of vertical stratification, i.e. belonging to a single landscape zone.

Thus the south-western slopes of the Great Caucasus region with wet subtropical features must be differentiated into an independent Colchidian province classified with the zone of wet subtropics. The stratification is characterized by the following series: (1) a foothill stratum of colchidian-type forests of zheltozems and krasnozems; (2) the mid-range stratum of beech forests with evergreen subforests on brown, mountain-forest soils; (3) the mid-range stratum of dark-conifer forest, including evergreen scrub on brown mountain-forest soil; (4) the subalpine stratum of scrub and tall grass; (5) the alpine-meadow stratum; (6) the nival-glacial stratum. The same 'colchidian' type of stratification (but with slight local differences) is found on the north-western slope of the Small Caucasus region where another province, the Adzhar, is differentiated. Finally, yet another province, the Rion, coincides with the inter-montane basin region of the same name and has a lowland character.

All three of these provinces, regardless of the fact that they belong to different azonal regions, i.e. alpine and lowland, comprise in fact a single entity—the zone of Caucasian wet subtropics. (Fig. 47)

In cases where substantial differences in the structure of vertical stratification on opposing slopes, due to the effects of exposure, are found in an area of mountain range belonging to a single region and a single landscape zone, two independent provinces must be differentiated. An example is the forest-steppe zone of the Ural mountains, which occupies a major part of the southern Urals region. The stratification of the western and eastern slopes in this section of the Urals exhibits marked differences. In the east it is more continental in character and instead of broad-leaved and dark-conifer forest a stratum of birch and mixed deciduous-pine forests is found. These differences are further underscored by differences in the character of relief. In the west relief is more complex and absolute elevations are greater than in the east. For this reason, two provinces—western and eastern—can be distinguished in the forest-steppe section of the southern Urals. Obviously each of these provinces should be related to a different physical-geographic district; the former to the eastern-European district and the latter to the western-Siberian district. It can

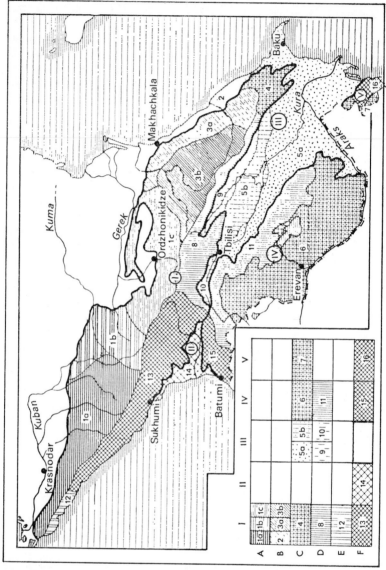

Fig. 47 Physical-geographic regionalization of the Caucasus (after A. G. Isachenko)
Landscape regions: *I*, Great Caucasus; *II*, western Transcaucasia; *III*, eastern Transcaucasia; *IV*, Armyan highlands and Little Caucasus; *V*, Talysh. *Landscape zones, provinces and subprovinces*: *A*, steppe zone: *1*, Ter-Kuban alpine

be seen from this example how alpine provinces need to be apportioned between different districts.

It follows that zonal physical-geographic classification reflects the connection between alpine and lowland landscapes. Naturally foothill landscapes are more closely and comprehensively bound up with the lowland landscapes of the corresponding zone than the mid-range and, in particular, the high-mountain landscapes. As a result, the higher the absolute altitude the less well defined are the zonal boundaries. The relationships between the higher taxonomic units of regionalization in mountain districts can be illustrated by the example of the Caucasus. (Fig. 47)[17]

There remains the problem of the lower taxonomic units in the regionalization of alpine districts. This question has not yet been sufficiently studied and no single solution has been proposed. The structure of vertical stratification within a landscape province is not entirely uniform and local variants exist owing to the altitude and the character of the orographic dissection of the range, e.g. the presence of a series of parallel ridges and valleys or lateral gullies with a specific orientation with respect to the prevailing winds, the presence of deep tectonic basins, etc. The Tyan-Shan and the Pamirs-Alai, for example, exhibit a characteristic alternation of high ranges, extending from west to east, and longitudinal valleys. Since these valleys are closed from the west, the air masses penetrate deeply into the alpine region and, as they rise, supply abundant rainfall to the slopes of the high mountain junctions (e.g. Matchin, Talgar, etc.) which enclose the

---

province (*1a*, Kuban alpine barrier foot subprovince, with bands of broad-leaved and dark-conifer forests; *1b*, El'bruss alpine-steppe subprovince; *1c*, Ter subprovince with a band of broad-leaved forests on outer slopes, with pine forests and highland xerophytes in the barrier shadow). *B*, semidesert zone: *2*, Caspian-Dagestan plain province; *3*, Dagestan alpine province (*3a*, outer Dagestan subprovince with alternate steppe and forest areas); *3b*, inner Dagestan subprovince with mountain steppe and highland xerophytes). *C*, desert zone: *4*, Shemakhin-Kobystan alpine desert-steppe province; *5*, eastern-Transcaucasian province (*5a*, Kura-Araks plain subprovince with deserts and piedmont steppes; *5b*, Shirak-Adzhinour low-alpine desert-steppe subprovince); *6*, desert-steppe Armyan highlands province; *7*, Galyshin alpine-steppe province. *D*, transitional zone of arid forests and cinnamonic soils: *8*, alpine province on southern slopes of the Great Caucasus, with a broad-leaved forest area; *9*, Kakhetin (Alazan-Agrichai) plain province; *10*, Gori plain (intermontane) province; *11*, Little Caucasus alpine province with an area of broad-leaved forests. *E*, mediterranean zone: *12*, Novorossiysk alpine province. *F*, fragments of wet-tropical zone (forest landscapes along barrier foot): *13*, Colchidian (Abkhaz-Svanets) alpine province; *14*, Rion plain province; *15*, Adzhar-Imeritin alpine province; *16*, Lenkoran low-alpine plain province.

valleys in the east. As a result, the snow level is lowered in upper valleys and the upper (glacial) stratum is much broader.

The northern slope of the Great Caucasus comprises in the western part (the Kuban province) numerous lateral spurs running in the north-north-east direction, which, together with the longitudinal valleys in the central watershed range, the lateral range and the cuesta series, form an entire series of valleys more or less unaffected by the prevailing wet winds. As a result, individual valleys exhibit particular climatic characteristics which are also reflected in the vertical stratification. Its features are more continental and progressively intensify in an easterly direction. Thus the dark-conifer forest stratum, well defined in the upper valley of the Great Laba, is no longer continuous further to the east in the Tevyerda valley, while in the upper section of the Kuban the stratum is entirely absent. The opposite is true of the pine-forest stratum; its thickness increases in an easterly direction. The determining factor is the increased altitude of the range in the same direction; this results in incomplete vertical stratification in various parts of the province. The glacial stratum in the western part of the province either does not exist or exists only in fragments, while in the east it is quite substantial.

A number of authors have discussed the need to differentiate parts of mountain ranges characterized by a specific local stratification spectra as independent regional units. V. B. Sochava designates these units as *okrugs* or *landscapes*, I. S. Shchukin as *rayons*, and V. M. Chetyrkin as *regional complexes*. In general, however, the most frequently used term, 'rayon', denotes something different for each author (this is true for both alpine and lowland regionalization, but is especially marked in alpine regionalization).

This discussion leads to the differentiation of three levels of regionalization within alpine provinces: *okrug, rayon and landscape*. Okrug, as proposed by by V. B. Sochava, includes that part of a mountain province which is characterized by a specific local variant of a given type of stratification. A rayon is a geomorphologically well-differentiated part of an okrug, definable within a single stage* of vertical stratification (low, central and upper) and sometimes within an individual inter-montane basin, valley, etc. The definition of a mountain landscape, together with appropriate examples, has already been provided in chapter 3. It will be recalled that a mountain landscape must be uniform with respect to its structure of vertical stratification, its position within a certain stage, and its structural-lithological classification.

* See chapter 3, pp. 162-9.

In the western part of the Colchidian province of the Great Caucasus, the Abkhaz okrug, which embraces an entire system of stages from the foothills to the crests, can be differentiated. The lower parts of that okrug are distinguished by relatively low moisture supplies, although on the whole the 'colchidian' type stratification prevails. A piedmont hill rayon consisting of Tertiary gravels, conglomerates and marls with a stratum of colchidian broad-leaved forests can be clearly differentiated; next, a mid-mountain range rayon, and finally a high-alpine rayon. In each of the rayons in turn landscapes can be distinguished. In the mid-mountain rayon, for example, highly distinct landscapes of limestone ranges with karst relief and a dominant stratum of beech forest on mountain soddy-carbonate soils, together with landscapes on ridges composed of Jurassic sandstone and porphyrites with beech and dark-conifer forests on mountain-forest brown soils, can be differentiated. Individual landscapes constitute deep longitudinal valleys in the Lower Jurassic clay shales along the boundary between the mid-mountain part of the okrug and the high-alpine region in the axial part of the Great Caucasus. In addition, the highest part of the mid-mountain rayon, with subalpine and alpine meadows, must be differentiated into independent landscapes.

Therefore, the taxonomic system in alpine regionalization is found to be more complex than the system accepted for the plains. This is due to the greater complexity of alpine regions. By contrast with the situation on the plains, in alpine regions the landscape and the physical-geographic rayon do not, in general, coincide. In certain specific conditions, however, they may do so. As noted earlier, small alpine 'islands' are regarded as individual okrugs. If, however, they are not very high and are not divided into rayons, they are regarded as a single rayon. Furthermore, in cases when they consist of a single structural-lithological complex (e.g. a volcanic hill on a lava base, a granitic intrusion, etc.), it may be sufficient to classify them as a single landscape. It follows from the above considerations that 'okrug' constitutes a level of regionalization which is obligatory only for medium-high and high-altitude mountains; in low-level mountains, 'okrug' and 'rayon' coincide.

The relationship between landscape regionalization and the regionalization of individual geographic components

In addition to landscape (multiple-component) regionalization, territories are also regionalized with respect to individual physical-geographic components: climate, relief, vegetation, etc. In view of the close relationships obtaining between these components, a close correlation can be expected between different types of regionalization. Many prominent Russian scientists have contributed to the concept of unity of natural regionalization, i.e. the conformity of all taxonomic systems concerned with the earth's surface. Thus K. D. Glinka wrote as early as 1922 that soil regions are simultaneously climatic, botanical and geomorphological regions, i.e. natural-historical regions in a broad sense. Somewhat later the prominent botanist-geographer Yu. D. Tzinzerling wrote that 'essentially there are no individual geobotanical regions and, as far as geobotanical regionalization is concerned, we are in fact dealing with the geobotanical description of natural landscape regions'.[17] This view is shared by S. Ya. Sokolov and other geobotanists. Recently V. B. Sochava has been a prominent supporter of this conclusion.

Experience with natural regionalization shows that given a synchronized study of the various types of natural regionalization by workers in appropriate disciplines; the results obtained, if not entirely in agreement, are at least very close. In cases, however, where investigations proceed without proper synchronization, using the data from a single component and taking no account of its geographic relationships, and employing different principles and methods, close conformity in the derived regionalization patterns cannot be expected. (These are essentially the reasons why such substantial differences exist between the results of the geomorphological, hydrological, geobotanical and natural-historical regionalizations of the U.S.S.R. in the various publications of the Committee on Natural-Historical Regionalization of the U.S.S.R. Academy of Sciences.) Understandably these studies provide no objective basis for conclusions regarding the conformity or non-conformity of the various types of natural regionalization.

There are no theoretical grounds for the rejection of the similarity between landscape regionalization and the branch-types of regionalization. Neither should we admit the idea that types of natural regionalization fully coincide with one another at every level of classification. *Correspondence* and *coincidence* are two different concepts.* Were it not so, the need for branch regionalization would disappear.

The spatial distribution of various geographic components is subject to general laws. Accordingly every natural boundary is essentially common to all components and has some specific significance within

---

* A discussion of correspondence and coincidence in relation to 'patterns of distribution' of landscape elements (components) is found in James (1934, p. 83; 1952, p. 217). In his 1934 article (p. 82) he admits that 'the terminology of this phase of geography is yet to be developed'. He goes on to suggest four types of pattern relationships, the terminology of which is somewhat refined in his 1952 article.

the landscape (i.e. it must be reflected in the landscape classification of terrain.)

The same boundary, however, may have a different taxonomic significance for various branches of regionalization. Thus certain common soil-geobotanical zonal boundaries may act as first-order boundaries for soil regionalization (i.e. dividing soil zones), but have a parallel second-order role in geobotanical regionalization (i.e. constituting merely the boundaries of subzones or bands). An example is the boundary between the podsolic and soddy-podsolic soil zones, or between the chernozems and chestnut soil zones.

Boundaries, the features of which are azonal, play a more important role in geomorphological regionalization than, for example, in geobotanical regionalization. By contrast, zonal boundaries valid especially for soil and geobotanical regionalization, are often secondary for geomorphological regionalization.

It follows that the taxonomic units of the different branches of regionalization need not coincide, and their level in the hierarchy may differ, even though every regional boundary must manifest itself in some way in every branch schema. Landscape regionalization, the character of which is the most complex and which is to an equal degree zonal and azonal, combines in itself all branch regionalizations. For this reason the units of landscape regionalization must correspond to other, not necessarily taxonomically equivalent, categories of climatic, geomorphological and soil regionalization, etc. (as shown earlier in the example of zones and subzones).

Many geographers have noted that passing from the higher to the lower taxonomic units, the conformity of the various types of natural regionalization gradually increases. The lowest level, i.e. the landscape, is essentially identical for all types of regionalization. The reason for this is that the landscape reflects the classification of a territory with respect to both zonal and azonal factors, i.e. it fully accounts for all factors in regional differentiation.

The territorial similarity between the landscape and branch regions is not always followed by the full conformity of all boundaries. As shown in the analysis of landscape boundaries (chapter 3), the spatial transitions are not equally distinct in every area. Even if they are mutually interdependent, this does not mean that they must always coincide. For example, climatic boundaries determined by relief cannot completely coincide with orographic lines. Since changes in climate are very gradual over an area and cannot be visually perceived, boundaries of climatic regions are more diffuse than, for example, the boundaries of geobotanical or geomorphological regions.

Something should be said at this stage about the relationships between physical-geographic and economic regionalization. These two types of regionalization are based on different sets of characteristics and, therefore, on different principles; for this reason there can be no direct relationship between them. Economic regions emerge, develop and transform much more rapidly than physical-geographic regions. They are related only in so far as economic activity is dependent on natural conditions—a very complex connection modified by social-historical conditions. Depending on these conditions, the effect of natural factors on the formation of economic regions varies. An example is provided by the history of the development of economic regions in the U.S.A.[18]

Even during the first half of the eighteenth century the foothills of the Appalachian mountains formed a considerable barrier to colonization and constituted the boundary of a large economic region. The Rockies, even though a much more substantial and difficult natural barrier, failed to retard colonization, which coincided with the mid-nineteenth century, and a much higher level of development of economic forces. The difference in economic development on either side of the Rockies is not nearly as great as that on either side of the Appalachian mountains, and the natural boundaries between the plains and the Rockies represented a characteristic 'axis' of an economic region. The most important economic centres are located precisely along that axis (e.g. in Denver, etc.).

The axis of another economic region (central-Atlantic) is defined by the so-called 'waterfall line', the most important physical-geographic boundary dividing the Atlantic plain from the Piedmont plain. Along this line, next to the sources of energy and natural harbours, the industrial centres of the central-Atlantic economic region, which includes both the Atlantic plain and the Piedmont plain, were developed.

Large rivers, as noted by L. Ya. Ziman, can also influence economic development and the formation of economic regions. Their usefulness in providing transport, water supplies and energy assists in the development of economic bonds between territories lying on both sides of a river. Very large rivers, on the other hand, may substantially hinder the development of economic interaction between such regions. For this reason, and depending on the historical situation, some rivers (e.g. the upper reaches of the Mississippi and the Ohio) constitute the boundaries of two different economic regions while others (e.g. the lower reaches of the Mississippi) constitute the axis of a single economic region.

Thus natural boundaries affect the formation of economic regions, and these effects can be correctly determined only by an analysis of the history of economic development, which is a function of economic geography.

Methods in physical-geographic regionalization

The acceptance of unified principles and a uniform system of taxonomic units is an important but certainly not sufficient condition for the attainment of homogeneous and entirely comparable results in regionalization. Success in this area depends to an equal degree on the methods employed and the degree to which the regionalization process is effectively organized. The combination of practical procedures and methods for the delineation and description of regional units which are capable of yielding homogeneous results consistent with the theoretical principles of regionalization is called the *method of regionalization*.

Until recently very little attention was devoted to the development of uniform methods of regionalization; every specialist proceeded on the basis of his personal experience, and of individual opinions about the best methods of evaluating the material and generalizing about it. Very often, these activities resulted, understandably, in the development of inadequate and unsatisfactory procedures.

Among these is the so-called method of overlay, i.e. the differentiation of regions by overlaying maps each expressing a particular branch of regionalization—geomorphological, soil, etc.—of a particular area. This purely formal procedure does not impose on its author the task of an independent analysis of the factors in geographic differentiation. The different details within each pattern, the different approaches of the author to the principles and methods of regionalization, the differences in the quality of original data, the lack of conformity among the 'branch' boundaries, etc., all combine to produce an *ad hoc* result. The acceptance of the overlay method is equivalent to the admission that physical-geographic regionalization cannot be based on an independent methodology. The complexity of regionalization must not be regarded simply as an eclectic mixture of components, each expressed by a single overlay. We therefore unhesitatingly reject this method.*

Often encountered in literature on regionalization is mention of the method of the *leading factor*. In practice, the leading factor comprises any one of the geographic components, and the method amounts to

---

* Overlay methods are currently used in town and country planning projects. (McHarg, 1969)

the differentiation of each taxonomic regionalization unit with respect to a single component (e.g. zones, according to the climatic indices for vegetation; provinces, according to relief; regions, according to lithology, etc.). It is obvious that, conceived in this way, the leading-factor method is essentially no different from the overlay method.[18]

The leading factors generally confuse two different concepts: factors or causes of physical-geographic differentiation which played the major role in the development of the given region, and the characteristic features which express most clearly the effect of the factors in regional differentiation. It is well known that the causes underlying the differentiation of regions are dual in character: zonal and azonal. The analysis of these, with reference to the genetic system, is the necessary prerequisite for any natural regionalization. Understood in this way, the leading-factor method may prove to be one of the principal methods in physical-geographic regionalization.

The features along which regional boundaries are defined are manifested to some degree in every component. These must be utilized as fully as possible without restriction to any preconceived component, so as to take into account the changes in surface relationships as well as the characteristic forms of topography, the distribution of lakes or marshes, the character of soils, the vegetation cover, and so on. It must be stressed that individual sections of regional boundaries may be most clearly defined in different components. In one place the boundary may follow a distinct faultline, in another a more or less distinct change in surface deposits accompanied by a change in soils and vegetation, in the third only a gradual change in the soil and vegetation cover due to the change in zonal conditions, and in a fourth the coastline, etc.

In isolating regional complexes the task is to establish the spatial relationships between the components, relying primarily on different types of maps (e.g. hypsometric, geological, soil, etc.) as well as the published statistical material and other data. In this case the method of *combined regional analysis of components* is employed. Field methods play the most important role in the differentiation of landscapes, but at higher levels of regionalization (the first and second order) these methods can only have a supplementary function.

In recent years increasing attention has been paid to the method of indicating regional complexes on typological landscape maps. By this method each successive regional unit of higher rank is differentiated on the basis of the distributional characteristics of more simple complexes. Thus landscapes are defined on the basis of typical combinations of urochishcha, and higher regional units on the basis of combinations of landscapes. In both cases the source is the landscape map (in the first case, essentially medium-scale maps; in the second, small-scale maps). This method is in fact nothing other than regionalization 'from below'. It has long been used in certain geographic disciplines including geobotany and soil science. Soil or geobotanical regions constitute specific territorial combinations of different soils or different plant communities respectively; they appear on typological maps, and for this reason such maps provide a very important and objective basis for regionalization. In landscape science this method did not gain acceptance until recently, only because landscape mapping is a very recent technique.

It would not, however, be correct to assume that regionalization 'from below', on the basis of a typological landscape map, is the only correct or universal method. It must be combined with the analysis 'from above' of the common factors in physical-geographic differentiation, relying on the existing, perhaps very approximate (first-order) systems of zonal and azonal classification. In the regionalization process this general system will be refined on lower levels. To differentiate, for example, landscape provinces, they must first be defined from above, relying on the general schema of districts, zones and regions, and analysing the reasons for their differentiation. At the same time, however, the differentiation of provinces must be approached from below, from concrete landscapes which are grouped in accordance with their genetic links. Without the latter, only an approximate system of provinces with rough boundaries would be obtained. The system is improved and made realistic by the approach from below.

Thus the principal methods in regionalization are the analysis of the zonal and azonal factors of regional differentiation and the analysis of typological landscape maps. Since very few such maps exist as yet, other methods are also in use, especially the combined cartographic analysis of components. All these methods, however, are essentially indirect. In any case, it must be recognized that regionalization is always a product of a combination of different methods. The use of any particular method depends on the required detail, the size of the territory, and the volume and the character of the data possessed.

The regionalization process consists of the following stages: selection of factual data, their analysis and synthesis according to requirements and consistent with theoretical principles, the differentiation of regions of different rank and the definition of their boundaries on maps, and the derivation of descriptions of the various regions.

Materials in physical-geographic regionalization comprise all data about the landscapes of the regionalized territory, including information on individual components. Cartographic materials are particularly important; they include a variety of specialized, as well as general-geographic (topographic) maps. The literature includes regional descriptions of a complex and branch nature as well as single contributions on more tangential questions. These are important for the evaluation of the natural characteristics of the given territory and its internal differences (e.g. contributions concerning the distribution of individual meteorological components, forest typology, characteristic geomorphological processes, etc.). In addition to printed contributions, existing unpublished material in the archives and libraries of scientific, educational and industrial institutions, e.g. diaries of expeditions, postgraduate dissertations, etc., should be taken into consideration.[19]

The first stage in the study of sources aims at explaining the position of the given territory in the system of macro-regionalization (i.e. in general first-order and, if possible, second-order physical-geographic regionalization) and the basic pattern of its zonal and azonal differentiation. It is self-evident that initially this requires a very careful study of the results of previous attempts to regionalize a given territory (assuming that they actually took place), and a study of regionalization systems of a more general character, including those which relate directly to the country as a whole. The results of this first stage of investigation are summarized in the form of a preliminary (working) regional system. Naturally it is hardly ever possible to make a system conclusive in any detail at this initial stage of regionalization.

The next stage consists of the classification and analysis of the material in literature and the statistical data on the outlined regions, aiming to produce a more accurate and detailed preliminary pattern and in particular to adapt it to the subsequent compilation of the description of every taxonomic unit.

The most important task in regionalization is the differentiation of landscapes. Regional units of higher rank (first and second order) in the territory of the U.S.S.R. are already relatively well defined and only local imperfections may need to be corrected. A key to greater precision in the entire system of higher regional complexes is the accurate differentiation of landscape. Because of the central position of landscapes in the system of geocomplexes, both field and laboratory methods are essential for their definition and investigation. First of all, however, map analysis, especially of landscape maps, must be carried out.

The differentiation of landscapes requires relatively detailed maps which reveal the morphological structure of landscape. Practical experience indicates that medium-scale landscape maps (approximately 1 : 1 000 000) are most suitable for this purpose. These still reveal, although in a very general way, the basic types of urochishcha. Since such maps are produced mainly by field surveys, landscape differentiation is, in the final analysis, a product of field investigation. In practice, landscape mapping and regionalization are carried out jointly by the same group of investigators. As a result, landscape boundaries are outlined in the course of field surveys and are more precisely defined later in geographic laboratories. Despite the fact that landscape boundaries are given a preliminary definition in field surveys, the results of field mapping must always be verified and supplemented in the laboratory with additional cartographic data and descriptive information from the literature, especially with regard to factors which cannot be directly studied by means of surveys, e.g. the climatic or hydrologic conditions. It should be remembered that landscape boundaries are often very diffuse and cannot always be accurately defined on maps showing the urochishcha.

This situation can be resolved by a cartographic analysis of the zonal and azonal factors in landscape differentiation and the subsequent analysis of landscape boundaries. For this purpose, maps of landscape-forming factors are compiled on the basis of existing branch maps (mainly of medium-scale). These include, first of all, a geological map, which shows the principal orographic lines, lithological bedrock complexes and Quaternary deposits, as well as certain characteristic genetic forms of relief, or their complexes (e.g. monticulate-morainic, karst, etc.). In addition, a map of climatic indices, which shows the zonal and provincial features of the climate and its dependence on relief, is compiled. Different indices may turn out to be relevant in each case. Most often they will include the mean temperatures of the hottest and the coldest month, the total sum of temperatures, the length of the vegetation period, and the annual rainfall.

Next, variations in the above factors are related to the distribution of soils and vegetation (additionally, a map of soil and vegetation cover can be compiled for this purpose). This enables fairly reliable definition of the most significant landscape boundaries; it is useful to transfer these to a separate map (see for example figure 19 in chapter 3). A comparative study of the results from such an analysis 'from above', with others established 'from below', i.e. by the study of the landscape map, provides the basis for the final subdivision of the territory into landscapes. The methods outlined above have been used in the physical-geographic regionalization of the north-western part of the Russian plain.[20]

Wherever landscape maps are not available, the significance of laboratory analysis of branch-type cartographic materials increases substantially. In these cases, the complexity of indices must gradually be increased first on intermediate cartographic diapositives (geological-geomorphological, climatic, soil-geobotanical), which are subsequently synthesized in a single map. On the first diapositive, geomorphological complexes are plotted and on the second, the complex soil-geo-botanical boundaries. Both are transferred to a single map using both shading (for geomorphological complexes) and a colour background (for the soil and vegetation cover). On the same map or, to avoid over-loading, on a copy are plotted the principal climatic and hydrological indices in the form of isolines, figures or symbols (e.g. to exhibit rain-fall). Furthermore, it is important in many cases to indicate certain boundaries having a special landscape-forming significance (e.g. boundaries of the last glaciation, the permafrost, etc.).

If we have at our disposal a sufficient volume of good quality and highly detailed cartographic materials concerning individual com-ponents, the geocomplexes of any given rank can be synthesized. At the present time, small-scale landscape maps which reflect the typo-logical classification of landscapes can only be produced for most of the U.S.S.R. by this indirect method. Medium-scale maps are very difficult to compile by purely laboratory techniques although, occasionally, depending on the quality of source materials, it is poss-ible to trace relatively clearly the principal types of urochishcha or groups of urochishcha (sites). The scope of this method is substantially broadened if aerial photography is used.

Wherever the territory under investigation has not been subject to special landscape studies and landscape surveys, field investigation must be carried out. The methods used in such investigations are a subject of a special study.[21] It is sufficient to note here that the volume and the character of field work in regionalization is determined by the level of knowledge about the given territory and the amount of detail required. Most of the territory of the U.S.S.R. has been studied in detail with respect to several geographic components, and the maps of these exist. In such cases it is not usually necessary to carry out the full range of field investigations, including a ground survey, etc. The volume of field work should be the minimum possible and should be limited only to inspection of key areas, aiming at the verification and refinement of results obtained in laboratory regionalization, the explanation of detected gaps, contradictions, etc. Inspection should only be carried out after all the data in the literature and cartographic sources have been considered, and after a preliminary differentiation of landscapes in the laboratory.*

For large areas of Siberia and the far east, however, we do not have at our disposal sufficiently detailed and comprehensive data for regionalization down to the level of landscapes. This task can hardly be accomplished without special investigations, including a large amount of field work.

Regionalization of the first and second order is based, as a rule, only on laboratory analysis; more precisely, on generalizations from field data available in the form of various maps, descriptions and statistical tables (e.g. on climate, hydrological conditions, and other numerical data). The criteria for the differentiation of higher regional units (districts, zones, etc.) has been discussed earlier.

Landscape survey maps which reflect the pattern of distribution of various types and species of landscapes provide the most useful basis for regionalization at higher levels. The differentiation of higher regional units must be based substantially on landscape typology. This does not mean, however, that landscape typology must always precede regionalization; the determination of classificatory groupings of landscape is greatly facilitated by the existence of general schemata for the zonal and azonal classification of continents. It follows that there is a mutual relationship between landscape classification and regionalization, and there is little point in asking which comes first. During investigation, both regional and typological classifications are rendered more precise and each 'benefits' from improvement in the other. It is important, therefore, to ensure that the full cycle of landscape analysis and systematization, including field survey, the classification of geocomplexes, the compilation of survey maps, and regionalization should be carried out by the same people.†

The results of regionalization are plotted on a map. Regionalization

---

* The methods of the CSIRO Division of Land Research are very similar: 'The time taken for a one-field-season survey is generally fifteen to eighteen months. The time taken for the various phases are set out below:
1. Pre-field-work—collection of existing information, preliminary aerial-photograph interpretation, selection of sampling site and planning of field itinerary, three months.
2. Field-work, three months.
3. Final aerial-photograph interpretation, three months.
4. Specialist evaluation of field data and co-ordination of land-system descriptions, three to four months.
5. Preparation of report, three to five months.
In some surveys, phases 3 and 4 are carried out simultaneously, but the total time taken is still the same.' (Christian and Stewart, 1968, p. 261)
† A similar approach is adopted by the CSIRO Division of Land Research (Christian and Stewart, 1968) and by the Soil Conservation Authority of Victoria (Gibbons and Downes, 1964), as well as by many other survey organizations.

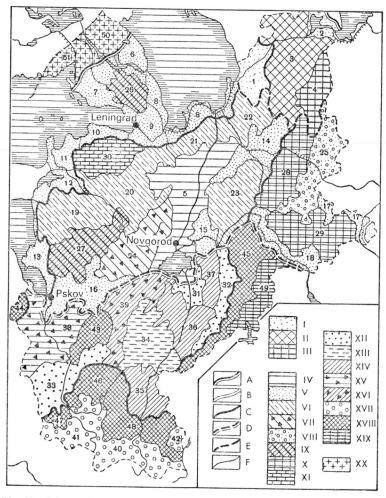

Fig. 48   Schema for physical-geographic regionalization and typology of land-scapes in the north-western landscape region of European U.S.S.R.

Physical-geographic boundaries: *A*, districts; *B*, regions; *C*, subregions; *D*, zones; *E*, subzones; *F*, landscapes. Arabic numerals designate individual land-scapes; roman numerals and hatchure indicate species of landscape.

Landscapes of the central-tayga subprovince of the north-western landscape region of the Russian plain: *I*, lowland lacustrine-glacial with light sediments (*1*, lower Svir; *2*, upper Svir). *II*, elevated hill kame-morainic (*3*, Svir-Oyat). *III*, elevated moraine-monticulate on limestone (*4*, Vepsov).

Landscapes of the southern-tayga subprovince of the north-western landscape region of the Russian plain: *IV*, low-lying lacustrine-glacial clayey (*5*, Ilmen-

maps do not have a complex character. Their content includes the boundaries of regional complexes of each rank, expressed by lines of different thickness, colour and type. To improve the clarity of maps, use can be made of appropriate backgrounds. It is generally useful to show zonal units (zones, subzones) by different colours and tints, and azonal units (districts, regions) by shading (mountain regions by dense shading). On maps which also include landscapes, the shading may express their typological similarity (i.e. landscapes of a single species are shown by the same kind of shading). These methods have been used on the regionalization maps of the north-west and the non-chernozem centre of the European part of the U.S.S.R.[22] Essentially, maps of this kind combine the typological and regional characterization of landscapes.

Figure 48 shows the regional subdivisions of various orders (down to landscapes) in the north-western part of the European U.S.S.R., within the Leningrad, Pskov and Novgorod regions. The boundaries of azonal units (districts, regions and subregions) are depicted by continuous lines; those of zonal units (zones and subzones) by discontinuous lines. The intersections of zonal and azonal boundaries yield the following units: landscape provinces produced by the boundaries of zones and regions, subprovinces produced by the boundaries of subzones and regions, and okrugs produced by the boundaries of sub-

---

Volkhov). *V*, low-lying lacustrine glacial, mainly with light sediments (*6*, lower Vuoksin; *7*, Primorsk; *8*, southern Ladoga; *9*, Neva; *10*, Pre-glint; *11*, lower Luga; *12*, Luga-Narva; *13*, Pskov-Chudo; *14*, Tikhvin; *15*, lower Msta; *16*, Pskov-Shelon; *17*, Mologa-Suda; *18*, central Msta). *VI*, low-lying morainic (*19*, Luga-Plyusa; *20*, Luga-Oredezh; *21*, Putilovo; *22*, Pasha-Syas; *23*, Msta-Vishera). *VII*, low-lying, on limestone moraine (*24*, upper Luga). *VIII*, elevated apron (*25*, Suda-Chagoda). *IX*, elevated monticulate kame-morainic (*26*, Lembolov; *27*, Pskov-huga). *X*, elevated morainic-monticulate on limestone (*28*, Tikhvin-Sherekhov; *29*, Kobozha-Uver). *XI*, elevated, on a limestone plateau, with thin till (*30*, Izhora).

Landscapes of the subtayga province of the north-western landscape region of the Russian plain: *XII*, low-lying lacustrine glacial, mainly with light sediments (*31*, lower Lovat; *32*, Pre-Valdai; *33*, Velikaya-Sorot). *XIII*, low-lying morainic-lacustrine-glacial marshy (*34*, Polista). *XIV*, low-lying morainic (*35*, upper Lovat; *36*, Polista-Lovat; *37*, Begla). *XV*, low-lying, mainly carbonate, lacustrine-glacial deposits (*38*, lower Velikaya). *XVI*, low-lying, on carbonate moraine (*39*, Shelon-bank). *XVII*, elevated apron (*40*, lower Velikaya-Lovat; *41*, Idritsa; *42*, Zhizhitsa). *XVIII*, elevated morainic-monticulate (*43*, Sudoma; *44*, Khanya; *45*, western Valdai; *46*, Kudevera; *47*, Sebezha; *48*, upper Loknya-Kunin). *XIX*, elevated morainic-monticulate, on limestone (*49*, eastern Valdai).

Landscapes of the southern-tayga subprovince of the Karelian landscape region (the Baltic shield district): *XX*, low-lying selga-trough, on Pre-Cambrian crystalline rocks (*50*, north-western Ladoga; *51*, Vyborg).

zones and subregions. The landscapes differentiated are grouped into species within landscape provinces and subprovinces.*

Where detailed landscape maps showing the distribution of the types of urochishcha exist, the boundaries of landscapes, provinces and other regional units can be plotted directly. The advantage of this method is that it not only enables the indication of regional boundaries, but also clearly reveals their content. The visual clarity in regional classification is lost, however, since regional boundaries affect the position of the second-level elements of the map (the major elements, represented by the coloured background, are the types of morpho- logical subclassifications of the landscape). At the same time, as noted earlier, such maps require a fairly large scale (not smaller than 1 : 1 000 000) and they are not really suitable for the expression of the regional classification of very large areas.

In general, very small scales are not required in representing the regionalization of a territory. In most cases the boundaries of individual landscapes may be shown schematically, even at a scale of the order of 1 : 10 000 000. Maps at this scale, however, are not easily read and are rather inconvenient for practical use. It should be kept in mind that by contrast with a typological map, where all boundary lines are grouped into a relatively small number of types, a regional map includes a large number of individual boundary lines, the number of which may reach several hundred for a single political-administrative or large physical-geographic unit. At a very small scale, even if every boundary line is distinctly separated, all the boundary lines together create a complex mosaic which is very difficult to read. The use of 'typological' colouring or shading substantially improves the visual clarity of the map without actually improving its utility.

The fact that regional maps are used for practical purposes makes it necessary to include within each region the principal topographic† objects (e.g. the drainage pattern, lakes, relief and townships, as well as political-administrative boundaries, etc.). In addition, a regional map needs a 'breathing space', i.e. sufficient space to introduce any kind of supplementary indices associated with its practical use. The most optimal scale, therefore, for the representation of a network of rayons (landscapes) is 1 : 1 500 000 to 1 : 2 500 000 on the plains and 1 : 1 000 000 to 1 : 1 500 000 in the mountains. To express results of

---

* For the most part an individual landscape is given the name of a river flowing through it.

† 'Topographic' is used here in an outmoded sense. 'The word "topography", before its misuse by geologists to mean landforms became common, was defined as the description of a small area'. (James, 1934, p. 85)

second-order regionalization, the scale of 1:5 000 000 or even less is quite sufficient, and for first-order regionalization, 1:10 000 000 and less.

In atlases regional maps usually have a smaller scale than component maps (e.g. geological, soil, etc.). Where, however, regional maps comprise a series characterizing different aspects of natural conditions, they are usually compiled at the same scale as other maps in the series. One example is the atlas of the *Natural Regionalization of Northern Kazahkstan*, published by the U.S.S.R. Academy of Sciences in 1960, in which the geomorphological, geobotanical and regional maps are all at the same scale.

The compilation of regional descriptions

The final stage of the regionalization process is the description of every regional unit. The structure and the content of regional description do not form part of a specific routine, but certain general principles should be observed. The principal requirement of regional description is that it should reflect the complexity of the system. Readers of such descriptions should have it clearly conveyed to them that a physical-geographic region is not simply a collection of various components within some *ad hoc* boundaries, but a regular and unique natural system.

Every regional description should begin with an introductory section enumerating the principles and methods of regionalization and providing an evaluation of the knowledge of the terrain, a discussion of basic data, and the history of regionalization in the area. The next stage should be a paleogeographic introduction, based on an analysis of the geomorphic history of the area, followed by an analysis of the contemporary factors in physical-geographic differentiation. A systematic review of each differentiated region, comprising the major part of the text, should then follow.

A description of each region should open with a discussion of its geographic position and its significance in the development of the physical-geographic features of the given geocomplex. The principal elements of these descriptions should also be the paleogeographic data, a discussion justifying the region's boundaries, and a review of its major components and their interrelationships. Special attention should be devoted to physical-geographic conditions which assist or hamper economic activity, health and recreation, etc., and also to un-utilized natural resources, as well as the need for protective or reclamation measures. There would be little point, however, in demanding from a landscape geographer detailed recommendations concerning the developments in economic activity,* agriculture, etc., since he is not adequately qualified and cannot assume responsibility properly belonging to experts in other disciplines.

Obviously the content of descriptions of regions of different orders will differ accordingly. A single formula could not even be applied to units of the same rank (e.g. provinces), since the nature of individual provinces differs widely. Even the problem of consistency in the analysis of individual components cannot be routinely resolved. In describing tayga landscapes, it would be pointless to attempt to analyse the podsol-forming process without giving a preliminary account of the typical features of the tayga vegetation. When describing the deserts, however, with their arid climate and sparse vegetation that results from the physical-chemical properties of the ground surface and affects the soil-forming process very little, it is more efficient to first review the soil-forming processes and later to deal with the vegetation.

A regional description proceeds from the general to the particular, i.e. from higher units to lower units. In characterizing the terminal, smallest units (e.g. provinces) repetition should be avoided, i.e. the duplication of the descriptions of preceding units (e.g. zones, districts, etc.). Instead the analysis should be deepened and rendered more concrete, mainly through a specific discussion of the given region.

Since there is no direct hierarchical dependence between zonal and azonal units, and since it is impossible to describe them together, we are faced with the problem of having to decide the best sequence for the description of regions of various categories and levels. There is no reason why this question should not be resolved arbitrarily, depending on the size and position of the region, the degree of zonal and azonal differentiation, and structural details.

First-order regionalization is restricted to the highest levels, and the most rational procedure here is to provide a brief review of all zones (together with subzones), then a review of all sectors and districts, and finally to conclude with a description of the zonal sections within the districts, i.e. zones in a narrow sense. Such a schema is well adapted to the territory of the U.S.S.R. with its distinct system of landscape zones, some of which extend across the entire area from the west to the east. For central and eastern Asia, however, which lack continuous zones extending from the western to the eastern boundaries, and where sharp contrasts exist between the monsoon eastern-Asian sector and the arid uplands of central Asia, it is more efficient to begin with a brief

---

* Many Western general-purpose survey organizations require of survey personnel fairly detailed recommendations for land use in their study area.

description of both sectors and separately describe each zone, before turning to districts and their zonal 'sections'.

In second- and third-order regionalization the descriptions may vary more widely but should conclude with an account of the units derived, i.e. either with provinces or subprovinces, or with landscapes (in alpine regions, we should conclude on the level of okrugs or rayons).

If the given territory conforms entirely to the boundaries of a physical-geographic district, the description should begin with a general discussion of the latter and only afterwards cover the principal features of constituent regions and, after that, zonal sections. Such a sequence will fully reveal the factors of geographic differentiation and will ensure a rational approach to the differentiation and description of landscape provinces and subprovinces at a subsequent stage.

Regionalization is often performed in the framework of political-administrative units or large economic units, which are somewhat heterogeneous in the physical-geographic sense. Thus the territory of the north-western economic region embraces parts of three physical-geographic districts—the Russian plain, the Baltic shield (Fenno-Scandia) and the Urals—and three zones—the tundra, tayga and subtayga. The method for the regional description of this area is as follows:

Following a brief account of the principles and methods of regionalization, a paleogeographic introduction is compiled, the purpose of which is to show how the process of historical development gave rise to the contemporary system of regions in that territory. In analysing the pre-Quaternary history, one concentrates on the azonal features in landscape differentiation and events which led to the formation of the most stable components of contemporary landscape, i.e. primarily the morphostructural elements of relief and the bedrock. The development of contemporary zonal differences is determined primarily by the Quaternary period; in reviewing this last stage of landscape history a more detailed analysis of the development of bioclimatic conditions must be presented.

The next chapter, still introductory in nature, should be devoted to an account of the principal features of the contemporary physical-geographic differentiation of the territory. It should not duplicate 'bit by bit' the traditional description of components. The problem is to reveal efficiently the most important contemporary processes which determine the internal differences, and those natural phenomena in which these differences are manifested. Naturally we should begin with an account of the distribution of solar radiation and the radiation budget. Next we should review the role of atmospheric circulation, the morphostructural peculiarities of relief affecting the distribution of heat and moisture. This is followed by a brief description of the major territorial differences with regard to temperature, moisture supplies, water budgets, runoff situation and degree of marshiness, differences in the contemporary geomorphological, geochemical and soil-forming processes, and in the character of vegetation cover and animal life. The description is concluded with a brief account of man's effect on the landscape, and of the level of economic utilization in relation to the natural features of the various parts of the entire territory.

Following the introductory chapters, the discussion of the regions proper is presented. In multi-level regionalization, the ultimate objective of which is to differentiate and describe landscapes, there is no need to characterize in detail, as is sometimes done, each of the established higher regional units, districts, zones, provinces, etc., as this leads to duplication in the presentation of data, unnecessary complexity, and overloading of the text. The most essential specific features of the main large regional units are already revealed by the paleogeographic introduction and subsequently rendered still more concrete in the discussion of contemporary differentiation. This discussion is designed to give a comparative account of the main features of landscape zones and subzones, districts and regions in terms of the above mentioned indices. This in turn makes special chapters devoted to zones, subzones, districts, etc. redundant.

Experience shows that a consistent regional analysis is conveniently carried out in terms of landscape regions, i.e. regions are selected as the basic intermediate stage in the transition from the general introductory description of a territory to the description of individual landscapes. A landscape region which groups together genetically similar landscapes constitutes the most appropriate background for a subsequent description of landscapes. Descriptions of regions include a number of different indices (e.g. the seasonal dynamics of landscapes) which are not easily provided for every individual landscape owing to insufficient factual material, or which need to be generalized over large territorial subunits so as to save space. Finally, it should be kept in mind that in accounting for the natural resources of large economic regions, not only a detailed discussion of landscapes is required but also a generalized picture in terms of relatively large natural territorial units. A landscape region satisfies all these requirements very well.

It follows that landscape regions should be described in somewhat more detail. Much more attention should be given to the description of the geology, which determines most directly the diversity of physical-geographic conditions and the contrasts between landscapes. Such a description is commenced by examining the characteristic features of Pre-Quaternary relief, together with the bedrock formation, and passes on to the discussion of Quaternary deposits and contemporary (sculptural) relief features. Brief descriptions of the climate, water conditions, soils, vegetation and animal populations are also provided.

A substantial portion of the text (approximately one-third) devoted to the description of a landscape region, is concerned with the description of the seasonal rhythm, a subject which only recently found its place in regional physical-geographic investigations. The seasonal rhythm reveals most effectively all the important structural features of landscape, the principal relationships between components, and the role of climatic factors in landscapes. The analysis of seasonal variations in the course of geographic processes is a necessary prerequisite for the evaluation of the conditions which govern the introduction of economic activity, especially agricultural activity. Naturally it would be extremely important to describe the seasonal rhythm of every landscape (as provided for in the programme outlined below) but this is often impossible owing to insufficient data. We accordingly restrict ourselves to a more generalized analysis within the framework of landscape regions, stressing at the same time the intra-regional differences and the specific features of individual typical landscapes. For each season and period (subseason) a consistent description is provided firstly of climatic processes and the characteristics of the principal meteorological elements, and later of hydrological and soil processes, floral and faunal phenological phenomena and, finally, the developmental phases of the major crops and the conditions governing agricultural production. The description is supplemented by tables of data which assist in a comparative analysis of the seasonal rhythm for a number of places within the given region.

Finally the description ends with a discussion of the major natural resources: power sources, mineral raw materials, forests, agro-climatic factors, etc.

A special section is devoted to the soils, which are described in terms of the predominating types of urochishcha and facies and with regard to their natural productivity and present-day condition.

The entire review of a landscape region is organized in such a way as to, on the one hand, emphasize its homogeneity and, on the other, reveal the major internal differences; in this way the groundwork is provided for the subsequent differentiation of a territory into provinces, subprovinces, okrugs and landscapes. The provinces and the subprovinces, however, are not described separately since they are sufficiently clearly differentiated with regard to relief, climate, seasonal rhythm, etc. in the descriptions of landscape regions. Only okrugs require very brief descriptions, essentially limited to an indication of the most specific natural features manifested in the character of the morphostructure, the history of landscapes, longitudinal-climatic characteristics, the level of cultivation, etc.

Landscape description occupies a central place in regional description. Even in cases where regionalization is confined, for some particular reason, to the higher levels of classification, the differentiation and description of these higher units (subprovinces, provinces, etc.) should be based on the description of landscapes.

In detailed regionalization of relatively small territories with only a few landscapes, each is given a separate description. Where, however, a large territory which includes tens or even hundreds of landscapes is regionalized, a relatively complete description would occupy too much space, perhaps several volumes. In such situations the reader finds it difficult to concentrate and to gain an impression of the major physical-geographic processes. The main effort should therefore be transferred to the description of groups of typologically similar landscapes, i.e. to a description in terms of species of landscapes (within each province or subprovince). Thus in the north-western southern-tayga subprovince of the Russian plain (Fig. 48) some thirty landscapes have been differentiated which can be grouped into six species (and two are further subdivided into subspecies). For example, the species of low-lying sandy lacustrine-accumulative southern-tayga landscapes includes thirteen concrete landscapes, and the species of low-lying morainic southern-tayga landscapes includes five concrete landscapes, etc. Cases exist where a species comprises only one or two landscapes within a given province. In the Pechora tayga province (Fig. 46) some thirty landscapes have been differentiated, most of which belong to seven or eight major species.

The structure and content of landscape description are determined by the nature of the landscapes, discussed in detail in chapter 3. The following programme is proposed for their description:
1. Geographic position (the description of the position of the landscape in the system of zonal-azonal classification and the analysis of its external links and boundaries).
2. The history of its development (in association with the paleogeography of the relevant province or region).
3. Contemporary zonal and azonal landscape-forming factors.
4. The principal features of individual components and their interaction.

5. Morphological structure (types of urochishcha and facies, their interrelationships and combinations, and the distribution of heat, moisture and solid material inside the landscape).
6. Seasonal dynamics.
7. General development trends, progressive elements and their influence on landscape development.
8. The effect of man's cultural activities on landscapes.
9. A discussion of natural resources, their rational utilization, protection and regeneration.

Landscape description should be supplemented with typical or generalized complex (landscape) profiles illustrating its morphological structure and the mutual relationships between the major components; fragments of landscape maps for the most typical (key) sections should also be provided.

At the contemporary level of knowledge about landscape in Russia, it is not yet possible to fully implement such a programme. It is all the more important, therefore, to make supporting use of the typological description of landscapes inside provinces, which enables the utilization of materials concerning different landscapes and mutually supporting one another. In addition, such an approach is very important in practice, since similar landscapes are uniform with respect to their productive potential and require similar economic, reclamation and protective measures.

It is necessary to indicate the specific individual features of each landscape, but this can be done in part within the framework of the typological description and in part by citing supplementary indices, mainly numerical. To save space and to facilitate comparison, these indices should be compiled in a tabular form. Tables should include the morphometric data (e.g. mean and maximum absolute surface elevation, predominant heights and slopes, etc.), climatic and hydrological characteristics (e.g. mean temperature during the hottest and coldest month, extreme temperatures, duration of the frost-free period, annual rainfall and its seasonal distribution, the value and the coefficient of runoff, the dates for freezing and thawing of rivers, etc.), as well as figures representing the level of development and the distribution of land use (e.g. level of agricultural activity, afforestation, level of marshiness, distribution of lakes, etc.). In describing the morphological structure of a landscape it is very important, provided appropriate material is available, to publish data on the approximate areal relationships between the major types of urochishcha.

5

Conclusion

The history of landscape science shows that its emergence and development was a direct result of practical need and accumulated experience. Man continuously encountered the objective existence of natural complexes and sooner or later had to realize his need for a complex account of local natural conditions as they related to agriculture. The first empirical attempts to derive such accounts were undertaken centuries ago (e.g. the official system for the qualitative evaluation of Russian soils during the fifteenth to seventeenth centuries—the so-called 'scribes books'), but the first scientific, geographic foundations for the resolution of this important national-economic problem were laid towards the end of the nineteenth century by V. V. Dokuchayev.

Dokuchayev showed that the correct planning of agricultural activity, anti-drought measures and measures against other unfavourable natural conditions, as well as the rational utilization and organization of a territory, must all be based on the study of geocomplexes, and in particular on the study of natural zones. Dokuchayev, therefore, developed natural-scientific principles for the conduct of agriculture and for a rational restructuring of a territory, and simultaneously founded a new science—landscape science, and contemporary general physical-geography.

The combination of practical needs with deep theoretical treatment of problems is typical of the work of all of Dokuchayev's followers. Dokuchayev's work stimulated the development, on landscape geographic principles, of such applied disciplines as silviculture, swamp-management and reclamation, pasture management and sand stabilization. The most prominent specialists in these areas, G. F. Morozov, G. N. Vysotski, R. I. Abolin, S. S. Neustruyev, B. B. Polynov, L. G. Ramyenski and others, made fundamental contributions to the theory of landscape science. During the very early stages

of landscape science G. F. Morozov expressed the view that silviculture and reclamation (in particular, agro-forest reclamation) constitute separate applied branches of geography, and that the training
of forest and reclamation officers would be most effectively carried out
in the departments of geography in the universities. It should be
remembered also that the first landscape map served the practical
problems of agriculture and reclamation.

Currently, ideas of landscape science are gradually penetrating into
the most diverse areas of national economy, science and culture. One
of the main areas of application in contemporary landscape science is
agro-landscape investigation.* The proper organization of agriculture
requires a complex and comprehensive analysis of all natural factors
including, apart from soil itself, the character of the locality (including
slope exposure, gradient, etc.), the microclimate, water supply, etc.
Very often the productivity of soils, the use to which they can be put,
and the extent to which their cultivation may be mechanized, depend
not only on soil quality but on many other factors as well. In the tayga,
for example, better-quality soils often lie in frost-threatened areas,
(i.e. in depressions); the best soils are often found on steep slopes
unsuitable for mechanized cultivation or generally unfit for agriculture
without special anti-erosion measures. An essential factor in agriculture is the landform features of areas, i.e. their size and the character of their interrelation and the degree of variety or uniformity (e.g.
the solonetz areas of varying size and the beech outliers on the upland
areas of the virgin steppe, the motley alternation of lands in monticulate-morainic landscapes, the complex landforms due to strong
gully dissection). It is obvious that the landform features are determined by the *morphological* factor of the landscape and must be taken
into account whether our concern is with developing virgin lands or
the rational utilization of cleared areas.

Thus, when we speak of 'lands' or 'land areas' (ugodye) we have in
mind a good deal more than the soil itself. An 'ugodye' is a *morphological unit* of the landscape (or the urochishche or facia) regarded
from the *agricultural point of view*.[1] The term 'ugodya' should be used
to refer not only to lands used in agriculture (cultivated ugodya) but
also to the potential soil resources (i.e. those natural ugodya which may
be brought in future under cultivation). It follows that a survey of
soils, their qualitative evaluation as well as the development of plans
for local land management and for the various reclamation and operational measures, must be based on a complex (landscape) investigation
and mapping of soils. The achievements of many groups of scientists

* (Zvorykin, 1963; Gvozdetskiy, 1964; Gedymin, 1968)

(including the Moscow, Lvov, Latvian and Estonian landscape scientists) have clearly shown the usefulness of this approach. Large-scale landscape mapping facilitates the fullest and most precise identification of areas with various agricultural properties, including the negative features (e.g. areas subject to outwash and erosion denudation, frost and drought). Landscape surveys provide, therefore, the most important scientific basis for the rational development of a territory.

Physical-geographic data enter into various areas of industrial, transport and communal planning although, as a rule, individual factors are isolated piecemeal for analysis and the overall complex is never fully considered. Interesting attempts have been made in recent years to apply the landscape approach employing landscape mapping and regionalization to the compilation of future plans for the development of cities and regions.

Yet another new and highly promising area of application for the theory of landscape has to do with the medical-geographic investigations and with resource-planning. It is well known that many of man's diseases are directly associated with the character of landscape—the animal world (the carriers of infectious disease agents), climate (a direct cause of many diseases),* geochemical conditions, the character of soils, waters, etc. The most recent investigations by medical geographers have shown that the explanation of the effects (both favourable and unfavourable) of natural conditions on human health must be based on the study of natural territorial complexes and, primarily, on the study of landscape as the principal territorial unit, and within it on the analysis of its morphological structure (in so far as the carriers of many infectious disease agents lie within certain specific uro-chishcha and facies). We already speak about the emergence of the *medical* landscape science.

We are also beginning to perceive ways of applying the theory of landscape to the study of radioactive emissions and their effect on living organisms and human health. The intensity of natural radiation, determined by the action of the cosmic rays and the presence of radioactive isotopes in the earth's crust, depends substantially on physical-geographic conditions. It reaches a maximum in high altitude mountain areas; it also varies within a broad range depending on latitude, the nature of rocks, water and soils. Tests of nuclear and thermonuclear weapons are a factor contaminating the atmosphere, and through it the soils, water supplies, plants and animals, with radio-

* (Critchfield, 1966)

active isotopes. Radioactive rain is distributed over the earth's surface unevenly, depending on atmospheric circulation, while conditions for the accumulation and migration of the atmospheric fallout depend on the character of the landscape—its water conditions, geochemical properties, soil character and plant cover. The lichens and mosses, for example, accumulate radioactive isotopes more intensely than other plants. The assimilation of these isotopes by humans and animals depends largely on the geochemical background. Thus in landscapes poor in calcium, its substitute, radioactive strontium-90, is assimilated particularly intensively, and this may lead to serious impairment of an organism's functions. It follows that different landscapes, urochishcha and facies differ with regard to their radiation level and this level (natural or artificial) should be monitored with respect to the landscape classification of the territory.

A promising development has been the emergence of landscape geochemistry as a branch of landscape science. Landscape-geochemical maps which reveal the different types of geocomplexes form a basis for the development of differentiated methods of exploration for certain mineral ores. Such maps may also find application in agriculture, where they may be used to indicate deficiences in mineral fertilizers, including trace elements, as well as in hygiene, sanitation and other areas.

Attempts have been made to produce surveys of underwater landscapes to assist in the scientific investigation of fishing areas. Detailed charts of underwater landscapes based on the study of relationships among the geographic components along the sea bottom—relief, soils, the character of the water strata and the biocenoses—provide bases for predictions about the distribution of shoals* in different seasons.

The theory of landscape constitutes the basis for the study of numerous natural phenomena (geological, permafrost, etc.). Its role increases substantially as aerial photographic methods of investigation gain wider acceptance in scientific and applied investigations. The concept of landscape and its morphological units represents the point of departure for any special (soil, hydrogeological, etc.) as well as topographic interpretation of aerial photographs. The interpretation of 'invisible' objects, e.g. soils, is only possible if all the various indirect physical-geographic features which are directly represented on the photograph (relief, vegetation, etc.), are considered. In the final analysis, photo-interpretation involves the differentiation of geographic complexes (frequently urochishcha) and the subsequent analysis of

---

* 'Shoals' is used here in the sense of schools of fish.

each complex with regard to soils and other elements. In fact, in ground-level mapping of soils we also mentally outline the geocomplexes in the field and plot the soil boundaries accordingly.*

The importance of landscape science in the life of society will increase immeasurably in the future. Under communism, the natural potential of every landscape should be developed most fully but in such a way as to ensure the maximum regeneration of renewable natural resources, mainly through increased biological productivity of the landscape. In cultural landscapes, natural processes will be directed towards the most appropriate objectives through the regulation of the energy and water budget, geochemical conditions, etc. Genuine cultural landscapes must ensure the optimum sanitation-hygienic conditions for human life and must also reflect the cultural-educational and aesthetic significance of nature. Finally, it is essential to preserve certain areas in their natural state so as to provide opportunity for the study of natural processes.

Of particular significance in the development of cultural landscapes is the problem of territorial organization, i.e. the restructuring of the morphology of the given landscape; the establishment of rational relations among the different types of land-use; the proper distribution of areas under crops, orchards, water reservoirs, plantations, populated areas, and industrial zones.

Thus an important task during the period of communist construction is the establishment of a theoretical base for the development of cultural landscapes. In the words of D. L. Armand, landscape science should transform itself into 'landscape development'. This grandiose concept elevates landscape science to a level on a par with the leading scientific disciplines, and places an obligation on landscape scientists to intensify their efforts both in the areas of theory and method of development and along the lines of closer association between theory and practice. The most pressing problems which await an early solution are reviewed briefly.

To begin with, ahead of any basic transformation of landscapes a complete analysis and a comprehensive qualitative evaluation of natural complexes on the basis of physical-geographic regionalization and landscape typology must be carried out. Regionalization and classification of geocomplexes is essential at all planning and design stages of measures aimed to utilize or transform a territory. Different branches of the economy have different and specific regional requirements. It follows that a single (universal) regionalization could not

* A similar approach to the classification of soil is reported in publications of the Military Engineering Experimental Establishment, Christchurch, England. (Beckett and Webster, 1965a, 1965b, 1965c; Brink *et al.*, 1966; Webster, 1965)

possibly satisfy all users unless it were supplemented from the point of view of individual practical need. Such regionalization constitutes the fundamental source for any applied purposes. The same is true of the classifications of geocomplexes. Various kinds of partial or applied landscape classifications are possible—physical-geographic, geo-chemical, engineering-geographic, etc. Each of these is based on a different set of criteria. For example, in classifying landscape in accordance with building construction requirements, it is necessary to begin by analysing relief (looking for flat areas), local material supplies, water resources, as well as climate and such specific factors as perma-frost, seismic character etc.

During the third Congress of the U.S.S.R. Geographic Society (1960) F. F. Davitaya offered the following example: The mid-mountain regions of eastern Caucasus and the Leningrad region have apparently nothing in common with regard to the natural conditions, yet from the point of view of potato cultivation they are very similar. By contrast, the natural characteristics of Livadia and Magarach along the southern Crimean coast appear to be identical and yet as far as viticulture is concerned they constitute quite different regions. Thus the groupings of landscapes with respect to the conditions for potato-cultivation will be very different from the classification in terms of viticulture. Before either of these classifications can be carried out, however, it is necessary to have at our disposal the necessary com-parative material and a record of all landscapes, each exhaustively described and systematized.

Given a sufficiently detailed and objective system of landscapes, the problem of designing any applied classification is reduced essentially to the selection and reorganization of material. It is not impossible that occasionally it may prove necessary to carry out supplementary field or laboratory investigations. Such investigations, however, require immeasurably less time and effort than that required when there is a total lack of a general, scientific system of landscapes (since in this case a new and comprehensive investigation would be required for each intended application).

It follows, therefore, that the problem of regionalization and classi-fication for applied purposes has two separate aspects: the develop-ment of a general, scientific genetic system of typological and regional classification of a territory, and the interpretation of this system from the point of view of intended utilization.

Of the many facets of this problem it is fundamental to emphasize the significance of the classification of elementary geocomplexes, facies, which constitute the basis for a functional typology and for the evaluation of soils, etc. in a broad sense. Such practical and at the same time theoretical problems as the qualitative evaluation of agricultural lands, forest typology, the census of pastures, marshes and peat moors, are directly related to the study and classification of elementary geocomplexes. In fact, all these points constitute different facets of the same problem; they are all variations of the applied typology of geocomplexes. As yet, however, there exists no single theoretical basis for such a typology. The current task for landscape scientists is to carefully investigate and generalize from the valuable experience accumulated by forest-typologists, marsh experts and specialists in agricultural land-use in order to compile a comprehensive system of lands and to develop the principles for their classification and qualitative (ecological) evaluation.

It is necessary to assume that for a long time yet landscapes will remain the principal object of the physical-geographic investigation in the U.S.S.R. Landscape constitutes precisely that category of classification which will in the coming years determine the degree of detail for the classification of the entire country. There are thousands of landscapes in the territories of the U.S.S.R., and we are still a long way from a complete picture. The existing regionalization systems provide only a general classification (first and second order) and we are only now beginning to approach the differentiation of landscapes (regions) in those parts of our country which have been most thoroughly studied. With regard to the classification of landscapes, the first stage—the production of a landscape map at a scale of 1:4 000 000—has commenced.

The most important practical task of Soviet geographers, at least during the next decade, is the development of a physical-geographic regionalization system for the territories of the U.S.S.R. down to the level of landscapes and the compilation of a 'landscape cadaster', i.e. full and comprehensive description of landscape. The solution of this problem must involve the compilation of a landscape map at a scale of 1:1 000 000.

A system of landscapes will simultaneously provide the basis for the study and classification of elementary geocomplexes (and, therefore, for the evaluation of lands); this parallel work should be carried out on the basis of large-scale landscape surveys. With regard to the complex classification and qualitative evaluation of soils, however, we will continue to be limited to the most important agricultural and forest regions, and territories which are of special significance for future economic development.

A landscape cadaster will also provide an important source for the compilation of regional landscape monographs concerning large economic regions, republics and other parts of the U.S.S.R. The compilation of such monographs has always been an important part of physical geography, although until recently, neither their content nor their scientific and methodological level satisfied the contemporary requirements of geographic theory and national economy.

One of the most topical problems directly concerning landscape science is conservation. It is necessary to be concerned not only with the conservation of individual natural objects (animal species, plants) or sections of landscape but with the conservation of entire landscapes, particularly those whose resources are subject to intensive utilization. Conservation must be based on the landscape-scientific approach and must account for the fact that a significant change in any one natural component upsets the natural relationships in the geocomplex. Felling of forests, for example, leads not only to reduction in timber resources but also to a deterioration in water supplies, to erosion and loss of valuable animal species, to flooding in mountain streams, etc. All these facts are fairly well known, yet one rarely finds them taken into account. A good and effective economic utilization of a landscape is inseparable from its conservation, the regeneration of its resources, and the consideration of the sanitation-hygienic, cultural-aesthetic and scientific significance of nature. All these factors must be taken into account in pioneering virgin lands, in regional planning or large-scale hydroelectric construction, thereby saving ourselves effort and expense in the future in repairing the damage resulting from inadequate development.

The active participation of geographers in the planning of the most important nature-transforming activities is unthinkable without a considerable improvement in the methods of landscape investigations. Two major problems emerge in this area. Firstly, it is important to develop much more deeply the methodology of landscape mapping and intensify the effort leading to the compilation of landscape maps for different purposes (general-scientific and applied) and at various scales (from detailed maps of key sections to general maps of the entire country). Landscape maps are the basis for the classification of geocomplexes, a qualitative evaluation of soils, physical-geographic regionalization, and the solution of many other theoretical, methodological and applied problems in our science.

The second important problem is the development of field-station studies employing quantitative methods, including geophysical and geochemical testing, aimed at improving our knowledge of landscape structure and its dynamics. At the present time this is the least advanced aspect of landscape study. And yet, it is quite obvious that unless such investigations are carried out, landscape science cannot be established as a precise science capable of prediction.

The progress of landscape science will depend largely on its association with the adjacent scientific disciplines—geographical, geological and biological. But the implementation of the conclusions derived by landscape science is not wholly under the control of landscape scientists. For the practical utilization of the achievements of our science to be turned into reality, the level of geographic knowledge among our people must be raised. We still cling to the traditional view of geography as a 'cab-driver's science' and secondary schools contribute very little to the elimination of this concept. Secondary school curricula almost entirely lack the elements of contemporary scientific ideas in geography. Geography courses in high schools (in higher forms) should include the study of landscape geography and should aim at developing in students an integrated dialectical concept of nature. In addition, specialized geographic training is essential for agronomists, forestry officers, economists, leading farm workers, and all those whose employment brings them into contact with the 'soil' and its utilization.

Notes and Bibliography

All Russian titles have been translated into English, but for the most part the papers are not available in translation. To establish whether English translations exist, it is suggested that the reader refer to the Israel Program for Scientific Translations Ltd, Jerusalem, or *Soviet Geography: review and translation* published by the American Geographical Society, New York.

Notes

1. SUBJECT, SCOPE AND HISTORY

[1] Kalyesnik, S. V. (1955), *Foundations of General Soil Science.*
[2] Solntsev, N. A. (1962), Major problems in Soviet landscape science, *Izvestiya Vsesoyuznogo geograficheskogo obshchestva*, **94** (1).
[3] Isachenko, A. G. (1961), *Physical-Geographic Mapping*, pt 3.
Vidina, A. A. (1962), *Methods in Large-Scale Field Investigations of Landscape*, Moscow: Moscow State University.
[4] Engels, F. (1940), *Dialectics of Nature*, London: Lawrence and Wishart, p. 215.
[5] Loc. cit.
[6] Ibid., p. 5.
[7] Ibid., p. 6.
[8] (1943), *Anti-Dühring*, London: Martin Lawrence, p. 27.
[9] Engels, F. (1940), op. cit., pp. 186-7.
[10] Mushketov, I. V. (1915), *Turkestan*. 2nd ed. vol. 1, pt 1, pp. 136-7.
[11] (1951), *Collected Works*, Moscow: Academy of Sciences of the U.S.S.R., vol. 6, pp. 388-9.
[12] Op. cit. (1949), vol. 1, p. 153.
[13] Op. cit. (1951), pp. 97-9.
[14] (1913), A proposed classification of Siberia and Turkestan into landscape and morphological regions, in *Collected Papers in Honour of Professor D. N. Anuchin's 70th Birthday*. Moscow.
(1915), *The Subject and the Aims of Geography*. Russian Geographic Society, **51** (9).
[15] (1910), *A Course in Physical Geography*, St Petersburg.
[16] (1914), A proposed epigenological classification of swamps, *Bolotovedene*, (3).

293

[17] (1938), *An Introduction to Complex Soil and Geobotanical Investigations.* Moscow: Selhozgiz.

[18] (1940), The tasks of geography and geographical field investigations, *Uchebnye zapiski Leningradskogo gosudarstvennogo universiteta*, (50).

[19] (1937), *An Analytical Characterisation of the Content and Structure of the Earth's Physical-Geographic Envelope*, Leningrad.

[20] (1948), The natural geographic landscape and some of its regular characteristics, *Trudy vtorogo vsesoyuznogo geograficheskogo sjezda*, **1**.

(1949), The morphology of the natural geographic landscape, *Problemy fizicheskoi geografii*, (16).

[21] Isachenko, A. G. (1964), The landscape map of the U.S.S.R., its scientific and practical significance. Materials for the 4th Congress of the U.S.S.R. Geographical Society, Symposium A, in *The Fundamental Problems of the Physical Geography of the U.S.S.R.*, Leningrad.

[22] (1931), *The Landscape-Geographic Zones of the U.S.S.R.*, pt 1, Moscow-Leningrad, p. 130.

[23] Gellert, I. (1959), *Geografiya ichosyaistvo*, (5).

[24] Kondracki, J. (1961), The state of the art in natural landscape in Poland, *Vestnik Moskovskogo gosudarstvennogo universiteta*, series 5 (Geography), **49** (4).

(1961), Types of natural landscapes in Poland, *Nauchnye Monografii Latviiskogo universiteta*, **37**

[25] Whittlesey, D. (1954), The regional concept and the regional method, in *American Geography*, edited P. James and C. Jones, Syracuse, New York: Syracuse University Press, p. 24.

[26] (1939), The nature of geography. *Annals of the Association of American Geographers*, **29** (3-4), pp. 425-6.

[27] Op. cit., p. 30.

[28] Ibid., p. 44.

2. PHYSICAL-GEOGRAPHIC DIFFERENTIATION

[1] (1962), Charts of atmospheric moisture content in the northern hemisphere, *Trudy pervoy nachnoy konferentsii po obshchey tsirkulyatsii atmosfery*, Moscow: Gidrometeoizdat.

[2] Evaporativity should not be confused with actual evaporation; the latter is always smaller since it is limited by the amount of actual rainfall. Evaporativity, on the other hand, is independent of the amount of rainfall since it is defined for conditions of unlimited moisture supply. In practice, evaporation from open-water surfaces is used as an evaporativity index, even though it differs somewhat in value from evaporation from vegetation-covered soil.

[3] Lvovich, M. I. (1961), The study of the water budget as one of the factors in landscape development, *Materials for the 5th All-Union Conference on Problems of Landscape Science*, Moscow.

[4] Lukashev, I. I. (1956), Zonality of geochemical processes in the weathering crust, *Izvestiya Vsesoyuznogo geograficheskogo obshchestva*, **88** (6).

Perelman, A. I. (1961), *Geochemistry of Landscape*, Moscow: Geografgiz.

[5] (1962), *Principles of the Theory of Lithogenesis*, Moscow: Academy of Sciences of the U.S.S.R.

[6] (1953), The climatic zonality of artesian waters (with reference to the U.S.S.R.), *Zapski: Leningradskogo gornogo instituta*, **29** (2).

[7] A detailed account appears in Isachenko, A. G. (1953), *Principal Problems of Physical Geography*.

[8] Ivanov, N. N. (1948), Landscape-climatic zones of the earth, *Zapiski Geograficheskogo obshchestva S.S.S.R.*, new series, **1**.

[9] Grigoryev, A. A. *et al.* (1956), The periodic law of geographic zonality, *Doklady Akademii nauk S.S.S.R.*, **110** (1).

(1960), The present state of the theory of geographic zonality, in *Soviet Geography*, Moscow: Geografgiz.

[10] (1950), Major trends in the development of a geographical environment, *Vestnik Moskovskogo gosudarstvennogo universiteta*, (3), p. 156.

[11] Strakhov, op. cit.

(1951), The age of the botanical regions in extra-tropical Eurasia, *Izvestiya Akademii nauk S.S.S.R.*, geography series, (2).

[13] Neyshtadt, M. I. (1957), *The History of Forests and Paleogeography of the U.S.S.R. in the Holocene*, Moscow: Academy of Sciences of the U.S.S.R.

[14] (1947), *Climate and Life*, Moscow: Geografgiz.

[15] (1958), *Collected Works*, 4th ed., vol. 38, p. 140.

[16] Ibid., p. 141.

[17] Ibid., p. 140.

[18] (1940), *Dialectics of Nature*. London: Lawrence and Wishart.

[19] Keller, B. A. (1923), *The Plant World of the Russian Steppes, Semideserts and Deserts*, no. 1, Voronezh, p. 56.

[20] For brevity, the strata are designated by their vegetational cover; the soil, climatic and other conditions also fully correspond to that cover.

[21] (1961), The altitude of localities, the age and structure of lowland landscapes, *Nauchnye monografii Latviiskogo universiteta*, **37**.

[22] (1909), Phyto-topological charts, their methods of compilation and practical value, *Pochvovedenie*, **2** (2), p. 111.

[23] Op. cit., p. 155.

[24] A special kind of zonality is manifested in areas where, as some authors allege, it does not exist at all, namely in alpine regions. In such regions zonality neither disappears nor is 'replaced' by vertical stratification but is merely transformed in characteristic azonal conditions.

[25] (1931), The geographic distribution of steppes and deserts with respect to to soils, *Zapiski Geograficheskogo obshchestva S.S.S.R.*, **5**, pp. 70-1.

[26] (1959), *The Physical Geography of Oceans and Continents*, Moscow: Moscow State University.

3. THE STUDY OF LANDSCAPE

[1] (1962), *The Morphological Structure of the Geographic Landscape*, p. 44.

[2] (1956), The principles of physical-geographic regionalization, p. 364.

[3] (1957), Some fundamental problems and tasks of Soviet physical geography, *Izvestiya Vsesoyuznogo geograficheskogo obshchestva*, **89** (2).

[4] (1952), The climate, macroclimate, local climate and microclimate, *Izvestiya Vsesoyuznogo geograficheskogo obshchestva*, **84** (3).

[5] (1959), The present state of landscape science, *Materials for the 3rd Conference of the U.S.S.R. Geographical Society*.

[6] Solntsev, N. A. (1960), The mutual interrelationships of 'living' and 'non-living' nature, *Vestnik Moskovskogo gosudarstvennogo universiteta*, (6).

[7] (1940), *Dialectics of Nature*, London: Lawrence and Wishart, p. 205.

[8] Ibid., p. 206.

[9] Loc. cit.

[10] Ibid., p. 318.

[11] Lavrenko, E. M. (1949), The phytogeosphere, *Problemy fizicheskoi geografii*, (15).

[12] (1955), *Foundations of General Soil Science*, p. 455.

[13] Kalyesnik, S. V. (1955), op. cit., p. 13.

[14] A detailed account is available in A. G. Isachenko (1960). The concept of 'type of locality' in physical geography, *Vestnik Leningradskogo gosudarstvennogo universiteta*, (12).

[15] (1961), Supplementary materials in landscape morphology, *Vestnik Moskovskogo gosudarstvennogo universiteta*, (3).

[16] (1956), The morphological structure of geographic landscape, *Izvestiya Vsesoyuznogo geograficheskogo obshchestva*, **88** (4).

[17] (1955), *The Study of Facies*, Moscow-Leningrad: Academy of Sciences of the U.S.S.R., p. 7.

[18] (1945), Facies, geographic aspects and geographic zones. *Izvestiya Vsesoyuznogo geograficheskogo obshchestva*, **77** (3), p. 162.
[19] (1956), *The Study of Landscape: selected papers.*
[20] Ibid., p. 498.
[21] Kalyesnik, S. V. (1955), op. cit., p. 24.
[22] Shchukin, I. S. *et al.* (1959), *The Life of Mountains*, Moscow: Geografgiz, p. 14.
[23] (1961), *Geochemistry of Landscape.*
[24] Rodin, L. E. *et al.* (1955), The exchange of ash elements and nitrogen in certain desert biogeocenoses, *Botanicheski journal*, (1).
[25] (1961), op. cit., p. 27.
[26] (1955), *Chemical Geography of Continental Waters*, Moscow: Geografgiz.
[27] The geochemical classification of landscapes constitutes a special problem with which we are not concerned here. (See Perelman, A. I. (1960), The geochemical principles of landscape classification, *Vestnik Moskovskogo gosudarstvennogo universiteta*, series 5 (Geography), 4.
The general classification of landscapes will be discussed in a later section.
[28] (1960), Total energy loss in soil formation, with reference to hydrothermal conditions, in *Heat and Water Conditions in the Earth's Crust*, Leningrad: Gidrometeoizdat.
[29] The method of calculation has been described in (1953) *Principal Problems of Physical Geography.*
[30] (1960), The role of phenological observations in intra-landscape regionalization, *Trudy fenologicheskogo soveshchanya*, Leningrad: Gidrometeoizdat.
[31] (1947), *Geographic Zones of the U.S.S.R.*, p. 21.
[32] Ibid., pp. 21, 23.
[33] (1942), The concept of growth in phytocenology, *Sovetskaya botanika*, (1-2).
[34] (1925), Landscape and soil, *Priroda*, (1).
[35] Op. cit., ch. 1, footnote 21.
[36] F. Engels, op. cit., p. 292.
[37] Ibid., p. 295.
[38] Ibid., p. 294.
[39] (1955), The origin and types of natural boundaries, *Izvestiya Vsesoyuznogo geograficheskogo obshchestva*, **87** (3), p. 275.
[40] (1952), Problems in complex physical-geographic investigations of city-development, *Problemy fizicheskoi geografii*, (28).
[41] Until recently, attempts to establish the types of vertical stratification in the U.S.S.R. were based exclusively on soil and geobotanical considerations. See V. M. Fridland (1951), Experimental soil-geographical classification of alpine regions in the U.S.S.R., *Pochvovedenie* (9); K. V. Stanyukovich (1955), The principal types of stratification in the U.S.S.R. alpine regions, *Izvestiya Vsesoyuznogo geograficheskogo obshchestva*, **87** (3).
[42] We exclude the landscapes of the Byleorussian and Ukrainian Polyessie which should be regarded as a separate subtype, transitional to western-European forest and western-European forest-steppe landscapes.

4. PHYSICAL-GEOGRAPHIC REGIONALIZATION

[1] (1947), *The Natural-Historical Regionalization of the U.S.S.R.*, Moscow: Academy of Sciences of the U.S.S.R., pp. 5, 21.
[2] Owing to war-time conditions, the results were published only during 1947-8.
[3] (1946), *Problems in Physical-Geographic Regionalization*, All-Union Geographic Society, **78** (1).
[4] (1960), The history of physical-geographic regionalization of European U.S.S.R., in *Physical-Geographic Regionalization of the U.S.S.R.*
[5] (1961), The relations and the consistency of typological and regional analysis of a territory, *Nauchnye monografii Latviiskogo universiteta*, **37**.
[6] (1962), *The Natural Regionalization of the Far East*, Irkutsk, p. 2.

[7] (1962), *Soil-Geographic Regionalization of the U.S.S.R.*, Moscow: Academy of Sciences of the U.S.S.R.
[8] (1956), *A Physical-Geographic Region and its Composition*, Moscow: Geografgiz.
[9] (1947), *Geobotanical Regionalization of the U.S.S.R.*, Moscow-Leningrad: Academy of Sciences of the U.S.S.R.
[10] Levina, F. Ya. (1959), The problem of zonality and the classification of European semideserts, *Botanicheskii journal*, **44** (8).
[11] Op. cit., footnote 3.
[12] A. A. Grigoryev isolates sectors within latitudinal belts, but these are not belts in our sense.
[13] Detailed account in A. G. Isachenko (1962), The basic principles of physical-geographic regionalization and problems associated with a taxonomic system of units, *Uchebnye zapiski Leningradskogo gosudarstvennogo universiteta*, **317**.
[14] Op. cit., ch. 3, footnote 3.
[15] (1959), Types of natural regions and boundaries, *Izvestiya nauk S.S.S.R.*, geography series, (6).
[16] (1961), Natural regionalization of the U.S.S.R., *Izvestiya nauk S.S.S.R.*, geography series, (3).
[17] Tsinzerling, Yu. D. (1932), The geography of the vegetation cover of north-western European U.S.S.R., *Trudy geomorphologicheskogo instituta Akademii nauk S.S.S.R.*, (4), p. 296.
[18] Ziman, L. Ya. (1948), Natural boundaries and the borders of economic regions, *Problemy fizicheskoi geografii,* (8).
[19] Certain general principles in the selection of materials and in work with source-data are discussed in N. I. Mikhailov, *Physical-Geographic Regionalization*, pt 3, Moscow: Moscow State University. For a discussion of special cartographic materials see A. G. Iaschenko (1960), *Physical-Geographic Mapping*, pt 2, Leningrad: Leningrad State University.
[20] Isachenko, A. G. (1961), Methodological problems in the physical-geographic regionalization of NW. Russian plain, *Problemy fizicheskoi geografii*, (55).
[21] Isachenko, A. G. (1961), *Physical-Geographic Mapping*, pt 3.
[22] (1961), Natural and agricultural regionalization of the U.S.S.R., *Problemy fizicheskoi geografii*, (55).

5. CONCLUSION

[1] Ugodya, defined in the study of the national economy, may also occupy individual parts of urochishcha; in such cases they constitute the latters' anthropogenic modifications.

Bibliography

Berg, L. S. (1947). *Landscape-Geographic Zones of the U.S.S.R.* Moscow: Geografizdat.

Isachenko, A. G. (1953). *Principal Problems of Physical Geography.* Leningrad: Leningrad State University.

—— (1961). *Physical-Geographic Mapping*, pt 3. Leningrad: Leningrad State University.

Kalyesnik, S. V. (1955). *Foundations of General Earth Science.* 2nd ed, Moscow: Uchpedgiz.

Milkov, F. N. (1959). *Principal Problems of Physical Geography.* Voronezh: Voronezh State University.

Natural and agricultural regionalization of the U.S.S.R. (1961). *Problemy Fizicheskoi Geografii.* (55).

Perelman, A. I. (1961). *Geochemistry of Landscape.* Moscow: Geografiz.

Physical-Geographic Regionalization of the U.S.S.R. (1960). Moscow: Moscow State University.

Polynov, B. B. (1956). *The Study of Landscape: selected papers.* Moscow: Academy of Sciences of the U.S.S.R.

Sochava, V. B. (1956). The principles of physical-geographic regionalization, in *Problems of Geography* (Collected papers of the 18th International Congress of Geography). Moscow-Leningrad: Academy of Sciences of the U.S.S.R.

Solntsev, N. A. (ed.), (1962). *The Morphological Structure of the Geographic Landscape.* Moscow: Moscow State University.

Zabyelin, I. M. (1959). *Theory of Physical Geography.* Moscow: Geografiz.

Editor's Bibliography

Armand, A. D. (1969). Natural complexes as self-regulating information systems. *Soviet Geography: review and translation.* **10** (1), 1-13.

Beckett, P. H. T. and Webster, R. (1965a). *A Classification System for Terrain.* Report No. 872. Christchurch, England: Military Engineering Experimental Establishment.

────── (1965b). *Field Trials of a Terrain Classification System: organisation and methods.* Report No. 873. Christchurch, England: Military Engineering Experimental Establishment.

────── (1965c). *Field Trials of a Terrain Classification System: statistical procedure.* Report No. 940. Christchurch, England: Military Engineering Experimental Establishment.

Bourne, R. (1931). Regional survey. *Oxford Forestry Memoirs.* **13**, 7-62.

Brink, A. B. *et al.* (1966). *Report of the Working Group on Land Classification*, pp. 3-16. Christchurch, England: Military Engineering Experimental Establishment.

Bunge, W. (1966a). *Theoretical Geography.* Lund: C. W. K. Gleerup.

────── (1966b). Locations are not unique. *Annals of the Association of American Geographers.* **56**, 375-6.

────── (1966c). Gerrymandering, geography and grouping. *Geographical Review.* **55**, 256-63.

Christian, C. S. and Stewart, G. A. (1952). *General Report on Survey of Katherine-Darwin Region, 1946*, pp. 5-24. Melbourne: CSIRO.

────── (1968). Methodology of integrated surveys, in *Aerial Surveys and Integrated Studies.* Paris: UNESCO.

Critchfield, H. J. (1966). *General Climatology*, pp. 362-6. Englewood Cliffs, New Jersey: Prentice Hall.

Crowe, P. R. (1938). On progress in geography. *Scottish Geographical Magazine.* **54**, 10.

299

Davis, C. M. (1969). *A Study of the Land Type.* Durham, North Carolina: U.S. Army Research Office.

Dickinson, R. E. (1970). *Regional Ecology.* New York: John Wiley.

Dunbar, C. O. and Rodgers, J. (1963). *Principles of Stratigraphy.* New York: John Wiley.

Gedymin, A. V. (1968). The use of old Russian land survey data in geographic research for agricultural purposes. *Soviet Geography: review and translation.* **9** (7), 602-24.

Geiger, R. (1966). *The Climate Near the Ground.* Cambridge, Mass.: Harvard University Press.

Gibbons, F. R. and Downes, R. G. (1964). *A Study of the Land in South-western Victoria.* Melbourne: Soil Conservation Authority.

Gould, L. H. (1967). *Marxist Glossary and Philosophical Dictionary,* p. 31. Sydney Current Book Distributors.

Grigg, D. (1965). The logic of regional systems. *Annals of the Association of American Geographers.* **55**, 465-91.

—— (1967). Regions, models, and classes, in R. J. Chorley and P. Haggett (eds), *Models in Geography,* p. 484. London: Methuen.

Gvozdetskiy, N. A. (1964). Physical-geographic regionalization for agricultural purposes. *Soviet Geography: review and translation.* **5** (7), 3-10.

Hartshorne, R. (1939). The nature of geography. *Annals of the Association of American Geographers.* **29** (3-4).

Herbertson, A. J. (1905). The major natural regions: an essay in systematic geography. *Geographical Journal.* **25**, 300-12.

—— (1912). The thermal regions of the globe. *Geographical Journal.* **40**, 518-32.

—— (1913a). Natural regions. *The Geographical Teacher.* **7**, 158-63.

—— (1913b). The higher units. *Scientia.* **14**, 199-212.

—— (1915-16). Regional environment, heredity and consciousness. *The Geographical Teacher.* **8**, 147-53.

Hills, G. A. (1960). Regional site research. *Forestry Chronicle.* **36**, 401-23.

—— (1961). *The Ecological Basis for Land-use Planning.* Ontario: Department of Lands and Forests.

James, P. E. (1934). The terminology of regional description. *Annals of the Association of American Geographers.* **24**, 78-92.

—— (1952). Toward a further understanding of the regional concept. *Annals of the Association of American Geographers.* **42**, 195-222.

Klingebiel, A. A. and Montgomery, P. H. (1961). Land capability classification, in U.S. Department of Agriculture, *Agriculture Handbook*, no. 210. Washington: U.S. Government Printer.

Kondracki, J. (1956). Natural regions of Poland. *Przeglad Geograficzny*. **27**, 48-60.

Linton, D. L. (1951). The delimitation of morphological regions, in Stamp, L. D. and Wooldridge, S. W. (eds), *London Essays in Geography*, pp. 199-217. London: Longmans, Green.

McHarg, I. L. (1969). *Design with Nature*. Garden City, New York: Natural History Press.

McIntosh, R. P. (1967). The continuum concept of vegetation. *Botanical Review*. **33**, 130-87.

Perelman, A. I. (1961). Geochemical principles of landscape classification. *Soviet Geography: review and translation*. **2** (3), 63-73.

Report on discussion of Anuchin's book (1961). *Soviet Geography: review and translation*. **2** (10), 18-32.

Rowe, J. S. (1962). Soil, site and land classification. *Forestry Chronicle*. **38**, 420-32.

Ruxton, B. P. (1968). Order and disorder in land, in G. A. Stewart (ed.), *Land Evaluation*. Melbourne: Macmillan.

Sauer, C. O. (1925). The morphology of landscape. *University of California Publications in Geography*. **2** (2), 19-53.

Saushkin, Yu. G. (1963). V. A. Anuchin's doctoral dissertation defense. *Soviet Geography: review and translation*. **4** (1), 53-9.

⸻ and Smirnov, A. M. (1970). Geosystems and geostructures. *Soviet Geography: review and translation*. **11** (3), 149-54.

Sibley, G. T. (1967). *A Study of the Land in the Grampians Area*. Melbourne: Soil Conservation Authority.

Soil-Geographic Zoning of the U.S.S.R. (1963), p. 13. Jerusalem: Israel Program for Scientific Translations.

Szava-Kovats, E. (1966). The present state of landscape theory and its main philosophical problems. *Soviet Geography: review and translation*. **7** (7), 28-40.

Tennessee Valley Authority, Hydraulic Data Branch (1965). *Design of a Hydrologic Condition Survey using Factor Analysis*, pp. 10-22. Knoxville.

Thornbury, W. D. (1964). *Principles of Geomorphology*. New York: John Wiley.

Unstead, J. F. (1916). A synthetic method of determining geographical regions. *Geographical Journal*. **58**, 230-49.

⸻ (1926). Geographical regions illustrated by reference to the Iberian Peninsula. *Scottish Geographical Magazine*. **42**, 159-70.

⸻ (1932). The Lötschental: a regional study: *Geographical Journal*. **79**, 298-317.

——— (1933). A system of regional geography. *Geography.* **18,** 175-87.

Van Riper, J. E. (1962). *Man's Physical World.* New York: McGraw Hill.

Webster, R. (1965). An Analysis of the Physiographic Basis of Soil Surveys. Ph.D. thesis, University of Oxford, Oxford.

Yefremov, Yu. K. (1961). The concept of landscape and landscapes of different orders. *Soviet Geography: review and translation.* **2** (10), 32-43.

Zvorykin, K. V. (1963). Scientific principles for an agro-production classification of lands. *Soviet Geography: review and translation.* **4** (9), 3-10.

Index

by Dorothy F. Prescott